Guzmán Mendoza Rojas
Ader Disney Román Neira

WASTE ROCK DUMP BLOCKS AND SUGAR CANE ASH

Guzmán Mendoza Rojas
Ader Disney Román Neira

WASTE ROCK DUMP BLOCKS AND SUGAR CANE ASH

A SUSTAINABLE ALTERNATIVE

ScienciaScripts

Imprint
Any brand names and product names mentioned in this book are subject to trademark, brand or patent protection and are trademarks or registered trademarks of their respective holders. The use of brand names, product names, common names, trade names, product descriptions etc. even without a particular marking in this work is in no way to be construed to mean that such names may be regarded as unrestricted in respect of trademark and brand protection legislation and could thus be used by anyone.

Cover image: www.ingimage.com

This book is a translation from the original published under ISBN 978-613-9-06958-3.

Publisher:
Sciencia Scripts
is a trademark of
Dodo Books Indian Ocean Ltd. and OmniScriptum S.R.L publishing group

120 High Road, East Finchley, London, N2 9ED, United Kingdom
Str. Armeneasca 28/1, office 1, Chisinau MD-2012, Republic of Moldova, Europe
Managing Directors: Ieva Konstantinova, Victoria Ursu
info@omniscriptum.com

Printed at: see last page
ISBN: 978-620-8-38496-8

Contents

Dedication

Dedication to my parents: Gregorio and Juliana for being the important support, to my daughters Lucero and Sofia for being the motivation to fight for my goals.

Guzmán Mendoza Rojas

Dedication to my parents: Teódulo Román López, Teodolinda Neira Chinchay and brothers, to my wife: Carlita Rubio, my children: Sofia, Liam and My Mateo, for being the pillar of my education and motivation in my improvement.

Ader Disney Román Neira

Acknowledgement

I thank first of all the advisor, teachers who imparted their knowledge, guiding me to achieve my goal.

Guzmán Mendoza Rojas

I thank GOD for my health and for being my guide, my parents for giving me life, my siblings, wife and children for giving me the strength to continue improving myself, my teachers for the contribution of their knowledge and for fulfilling my proposed goals.

Ader Disney Román Neira

SUMMARY

In this research work, the objective was to determine the effects on the mechanical properties of concrete blocks with the addition of recycled concrete and sugar cane bagasse ash residues (RCBCACR) in Lambayeque, 2023. For this research, a standard sample was designed and percentages of 5%, 10% and 15% of RCBCACR were added in order to evaluate the improvements in the mechanical properties of the concrete blocks.

This study seeks to evaluate the behaviour of compressive, flexural and tensile strength with respect to the standard sample in a concrete block, the research is experimental type, dependent variable properties of the concrete block and independent variable RCBCACR, the sample population was 180 concrete blocks, as an instrument were laboratory formats to obtain results of the tests performed.

The results obtained were, obtaining an optimum result with 10% RCBCACR; being 53.79 kg/cm^2 of compressive strength vs 0% RCBCACR of 51.08 kg/cm^2 at 28 days, a flexural strength of 6.19 kg/cm^2 vs 0% RCBCACR (5.80 kg/cm^2) flexural strength; with a compression vs. flexural ratio of 11.51%, a tensile strength of 1.09 kg/cm^2 vs. 0% RCBCACR (1.03 kg/cm^2) tensile strength; with a compression vs. tensile ratio of 2.02%.

In conclusion, the addition of 10% RCBCACR gives better compressive and flexural strengths, complying with Standard E.070, being able to work as NP and P block, regarding the tensile strength, the concrete block units with 10% dosage do not resist tensile stresses.

Keywords: Blocks, concrete, brick, ash, confetti, recycled, cane, bagasse, sugar.

I.- INTRODUCTION

Over the years, concrete has become one of the materials of great interest in construction, because its workability allows its adaptation to different types of constructions where it is used, however, in many occasions it presents very deficient properties in terms of durability and cohesiveness (Zambrano et al., 2021).

The main shortcomings of concrete are related to its brittle nature; due to the low durability of its strength, causing future crack formation, low tensile strength and low deformation capacity, new design alternatives should be investigated to obtain quality concrete using new materials such as making use of wastes that benefit the properties of concrete (Nunton et al., 2022).

On the other hand, most of the steps to elaborate concrete blocks are not the correct ones, given that most of the materials used present characteristics that are not suitable for mix designs due to their sizes, absorption percentages and variable shape percentages, influencing the final workability of the mixes, obtaining a concrete with poor properties, which is not recommended (Liu et al., Utilization potential of aerated concrete block powder and clay brick powder from C&D waste, 2020).

Given this, the final quality of concrete blocks is low, which depends largely on poorly performed manufacturing steps related to poor knowledge of the study of aggregates, incorrect determination of the dosage of each component of the mixes, and imperfect processing of the blocks with respect to moulding, compaction and curing (Liu et al., 2021).

Currently, it is easy to confuse cheap concrete with an economical one, given that resisting certain loads does not mean that it becomes durable, so that over the years its maintenance will be costly; on the other hand, concrete becomes economical if it becomes resistant and durable in the long term, implying higher initial costs in its production (Camarena & Díaz, 2022).

In another context, the growing environmental concerns in the world force concrete producers to use available wastes instead of natural aggregate and cement sources, making recycled concrete a more suitable and sustainable alternative worldwide due to the large consumption of concrete as in China and USA countries is excessive and is also reflected in cement production which its production is based on 85.9 million MT of cement per year (Song et al., 2019).

In January 2022, sugar cane production reached more than 768 thousand tonnes with a percentage variation of -0.5% to the production obtained in 2021, with bagasse being the formation of fibrous residues after the exclusion of the juice of the canes themselves, where the burning of this component generates ash, becoming components that are of no use to companies (INEI, 2022).

Due to all this, having a design and use of aggregates with low properties, the production of concrete blocks will have a short life span, notably losing its strength, and if these blocks do not have the right dimensions, they will not be placed quickly at the time of construction (Awoyera et al., 2021).

In addition, Peru has 1,813 authorised landfills (OEFA, 2023), leaving a significant percentage of waste without an authorised dump site. According to CAPECO, 70 % of construction waste goes to the sea, while 30 % is distributed at authorised points

(León, 2017).
Locally, in Chiclayo, 80% of concrete is deficient with respect to its properties when subjected to impacts, where if this same concrete is used for the production of concrete blocks, it will be a danger to users, due to its low resistance, in the construction of houses, these will be prone to collapse in a mild earthquake, if it were the case, generating economic losses and loss of life (Itamar et al., 2022).
Therefore, the need arises to give a new use to sugar cane bagasse ash waste, which, when incorporated into the production of concrete blocks, in addition to minimising the polluting impact that this generates, will directly benefit the users, as well as optimising the strength and durability properties, as well as a final economic benefit; by reducing the use of the other components that make up these mixtures (Anjos et al., 2020).
Furthermore, in reference to waste concrete, these are a viable option when making a good choice of the material to be used for the mix design, by means of sieving or other tests to determine its general characteristics, generating through its application an optimisation of the workability of the concrete, which leads to an increase in its strength (NedeljkoviC et al., 2021).
Therefore, this study seeks to optimise concrete blocks through the use of two new materials, which, although it is true that these components will be reused, will also obtain better results in the properties when they are added in the appropriate dosage percentages for the mixture, avoiding a decrease in its final properties.
Due to this situation, the following question is formulated: What are the effects on the mechanical properties of concrete blocks with addition of RCBCACR in Lambayeque, 2023?
This research is justified at an environmental level with the purpose of offering the reuse of materials such as RCBCACRs in order to make these components environmentally friendly, characterising concrete as a sustainable material, by making use of its waste, thus minimising the negative impact it may cause in the future.
For technical justification, a precise alternative is offered with respect to the significant characteristics that concrete blocks with the addition of RCBCACR can present; complying with the Peruvian regulations in force, guaranteeing the optimum use of these new eco-friendly materials.
In terms of social justification, the study is justified because it will provide an effective alternative for obtaining materials of excellent quality for the department of Lambayeque, which has a total of 1,197,260 inhabitants, where concrete blocks with additions of recycled materials will be used in various building structures in the future, acquiring significant benefits in the performance of their safety and functionality, being a benefit for users.
For economic justification, using these eco-sustainable materials made of concrete with added RCBCACR will reduce the production costs of concrete blocks, minimising the use of resources and reusing others.
Therefore, the following general objective is proposed: To determine the effects on the mechanical properties of concrete blocks with addition of RCBCACR in Lambayeque, 2023, and as specific objectives the following: i) To determine the optimal design of concrete blocks with addition of RCBCACR in Lambayeque, 2023.

ii) Determine the dosage in the concrete blocks with addition of RCBCACR in Lambayeque, 2023. iii) Determine the compressive strength of the concrete blocks with addition of RCBCACR in Lambayeque, 2023. iv) Determine the flexural strength of the concrete blocks with addition of RCBCACR in Lambayeque, 2023. v) Determine the tensile strength of the concrete blocks with addition of RCBCACR in Lambayeque, 2023.

By general hypothesis: The mechanical properties of concrete blocks with addition of RCBCACR increases significantly in Lambayeque, 2023. By specific hypothesis: i) The optimal design of concrete blocks with RCBCACR addition presents significant improvements in Lambayeque, 2023. ii) The optimal dosage in concrete blocks with RCBCACR addition presents significant improvements in Lambayeque, 2023. iii) The compressive strength of concrete blocks with RCBCACR addition increases significantly in Lambayeque, 2023. iv) The flexural strength of concrete blocks with RCBCACR addition increases significantly in Lambayeque, 2023. v) The tensile strength of concrete blocks with RCBCACR addition increases significantly in Lambayeque, 2023.

II.- THEORETICAL FRAMEWORK

At the international level, in the article *Sugarcane pulp sand and paper grain sand as partial fine aggregate replacement in environment-friendly concrete bricks* (Tayeh et al., 2021), the aim was to determine the influence of incorporating sugarcane pulp sand in concrete blocks. Samples with dimensions of 0.25 m χ 0.12 m χ 0.06 m were made. The results of the compressive strength test showed values of 26.53MPa, 27.2MPa, 26.9MPa, 25.53MPa and 23.87MPa respectively, as well as an absorption of 3.8%, 4.1%, 4.5%, 4.9% and 5.4% when sugar cane pulp sand was added at 0%, 10%, 20%, 30% and 40%. They concluded that by including 10% sugar cane pulp sand they achieved an increase in the strength of the concrete bricks and that all samples have water absorption values of less than 8%, complying with ASTM-C20.
In the article *Study on the properties variation of recycled concrete paving block containing multiple waste materials* (Wang et al., 2023), they aimed to replace natural aggregates with recycled concrete fine aggregate by including them in concrete paving blocks. They performed strength and absorption tests. They found that by using recycled concrete fine aggregates in proportions of 10, 20 and 30%, they achieved compressive strength values of 32.51MPa, 30.27MPa and 30MPa respectively, in addition to absorption results of 5.92%, 5.13% and 4.57% respectively. They concluded that 10% of fine aggregates of recycled concrete was the most suitable for the production of concrete pavers.
In the article *Sugarcane bagasse and rice husk ash pozzolans: Cement strength and corrosion effects when using saltwater* (Garrett et al., 2020), they aimed to evaluate the effect of partial cement replacements by including *sugarcane* bagasse ash and rice hus*k ash* in the mixes (Abasi et al., 2023). They used the corresponding tests to determine the required strengths and absorptions. They found that by using 0%, 10%, 20% and 30% of sugarcane bagasse ash and rice husk ash, they achieved a compressive strength of 18MPa, 25MPa, 17MPa and 18.5MPa for the first material and 20MPa, 25MPa and 24.5MPa for the second material, in addition, by absorption they reached values of 29%, 30.5%, 32.5% and 36% for the first material, and 30.8%, 32% and 405. They concluded that 10% of sugar cane bagasse ash was the most suitable as it had significant results, and 20% of rice husk was the most suitable for use in the production of concrete blocks.
In the article *An Experimental Study on Bricks by Partial Replacement of Bagasse Ash* (Prabhu et al., 2019), they aimed to conduct a study on the effects of bagasse ash addition in brick making. They carried out the corresponding tests using blocks with the addition of bagasse ash. The results showed that by using 6, 8, 16 and 20 % they achieved a compressive strength of 4.00N/mm^2 , 4.20N/mm^2 , 5.10N/mm^2 and 6.30N/mm^2 respectively, as well as an absorption of 20.20%, 19.80%, 19% and 17.90% respectively. They concluded that the inclusion of 20% bagasse ash had significant results in strength and absorption.
In the article entitled *The Utilization of Recycled Masonry Aggregate and Recycled EPS for Concrete Blocks for Mortarless Masonry* (Pavlu et al., 2019), they aimed to optimise the properties of concrete blocks by adding recycled masonry and recycled EPS. They carried out tests on the strength and absorption of the concrete blocks. The results of compressive strength when using 0, 35, 95 and 100 % of recycled

masonry showed values of 47.5MPa, 30MPa, 34MPa and 29MPa, and when using 10, 25 and 65 %, values of 13.5MPa, 21MPa and 18.5MPa for the same test, and for absorption they achieved 10.27%, 13.09%, 13.87% and 14.41% using recycled masonry and 12.07%, 11.62% and 12.92% when using recycled EPS. They concluded that the results were not adequate and recommended further tests with less than 35% addition of recycled masonry.

At the national level, Ccahuaya and Zeballos (2022) in their research *Improving the mechanical properties of the solid artisan wall by adding crushed recycled concrete for use in construction, Ilo, Moquegua, 2021* aimed to optimise the mechanical properties of each masonry unit by adding recycled concrete. They used a quasi-experimental design methodology, applied and quantitative approach. They achieved by results that by using 0, 10, 25 and 50 % of recycled concrete in each masonry unit they reached a compressive strength of 41.730kg/cm^2 , 75.540kg/cm^2 , 63.590kg/cm^2 and 60.060kg/cm^2 respectively, absorption of 3.650%, 2.370%, 2.650% and 2.800% respectively; as well as concave and convex warpage of 3mm - 1.3mm, 1.2mm - 0.9mm, 2.3mm - 0.1mm and 1.2mm - 1.1mm respectively. They concluded that the most optimum proportion of recycled concrete was 10% as it had significant results on their properties.

León and Reátegui (2020) in their research *Design of concrete blocks with the incorporation of sugar cane fibre for single-family houses in Moyobamba 2020* aimed to evaluate the influence of incorporating sugar cane fibres in concrete blocks. The methodology used was applied, experimental and quantitative. The results showed that adding 0, 15, 15, 17 and 19 % of sugar cane fibre to the concrete blocks had a compressive strength of 21,110, 18,930, 17,700 and 23,570 kg/cm^2 , absorption of 15,720, 13,230, 13,550 and 12.840 % respectively, and average warping with 19% addition obtained values on the bottom longitudinal and diagonal face of 1,310 and 1,400 mm and top longitudinal and diagonal face of 1,420 and 1,720 mm. They concluded that the most optimal dosage of sugar cane fibres was 19% as it generated an increase in compression.

Ardiles (2021) in his research *Influence of sugar cane bagasse ash as a partial substitute for portland cement type I in the production of masonry units Abancay, 2021* had the objective of finding out how sugar cane bagasse ash influences the partial substitution of cement in the production of masonry units. The methodology used was applied and experimental. The results showed that by using 0, 5, 10 and 15 % he achieved a compressive strength of 28.180, 33.590, 37.810 and 35.450 kg/cm^2 , absorption of 6.40, 6.20, 6.20, 6.20 and 5.70 % and in warping by using 15 % of ash had an average in its concave and convex lower face of 1.190 and 0 mm and concave and convex upper face of 2.650 and 0.700 mm. He concluded that the most optimal dosage was 10% as it had a better value in terms of compressive strength.

Correa and Polo (2019) in their research *Influencia de reemplazo de ceniza de caña de azúcar sobre las propiedades físicas y mecánicas de adoquines tipo II para pavimentos de tránsito ligero, Trujillo 2019* had the objective of finding out how the use of sugar cane ash influences the physical and mechanical properties of type II pavers. The methodology used was experimental and quantitative. The results showed that the addition of 0, 3, 6, 9, 9, 12 and 15 % achieved a compressive strength of 424,050, 447,160, 489,200, 506,960, 516,330 and 403,310 kg/cm^2 respectively, average absorption of 4,420, 4,230, 4,060, 3,260, 3,060 and 2,960 % respectively. They concluded that the most optimal dosage of sugar cane ash was 12% by generating an increase in compression and making the concrete impermeable.

Arope (2019) in his research *Compressive strength of concrete bricks f'c=210 kg/cm^2 , replacing the coarse aggregate by brick and recycled concrete, in different percentages* aimed to know how concrete and recycled bricks influence the compressive strength of concrete bricks. It used an applied experimental methodology. He achieved by results that, by using recycled concrete in 0, 10, 15 and 20 % had a compressive strength of 100, 105.060, 103.940 and 102.720 kg/cm^2 respectively, and by using recycled brick in 10, 15 and 20 % reached a compressive strength of 104.990, 104.440 and 93.610 kg/cm^2 He concluded that the most optimal dosage was 10% in both cases, since it increased the strength of the concrete bricks.

In the present research there are several theories related to the topic: According to Wu et al. (2022), sugarcane belongs to the group of tropical and subtropical crop belonging to the sugar crops in the world, when the juice of this same product is extracted, about 50% remains as bagasse. On the other hand, Almeida et al. (2019) mentions that mostly this component is burned as fuel in a combined generator to produce electricity, with sugarcane bagasse ash being the final waste material in the sugar production chain where about one tonne of bagasse can produce 25 to 40 kg of ash.

For Lyra et al. (2019) the ashes of this same material become a residual waste from the sugar industry, where this component is mostly burned as fuel, Garret et al. (2020) state that they are rich in amorphous silica that manages to be able to accommodate as one of the pozzolanic materials in concrete, in addition it is made up of calcium oxide in 6.83%, silicon oxide in 71.36%, aluminium oxide in 11.2%, iron oxide in 3.795 and magnesium oxide in 1.56%.

According to Kolawole et al. (2021), it is a material discarded by the sugar industries with a high content of pozzolanic components, these become produced from the by-product due to the extraction of the liquid from the sugar cane stalks which is the sugar cane bagasse, which is used by the industries as a type of fuel where the final residue is the ash of this by-product.

On the other hand, Khatab et al. (2021) state that recycled concrete is reused as aggregate to make new concrete, much of which can come from various sources, such as crushed concrete elements from buildings and other structures.

For Yuan et al. (2023) this component is derived from the remains of rejected units in precast concrete plants or from correspondingly tested concrete specimens, for this reason, the properties of recycled concrete usually vary depending on how they were made.

In addition, Carrico et al. (2021) mention concrete waste, which can be found as a type of waste due to demolitions, these wastes can present diverse properties when studied, because they present different types of compositions, due to the mixture, construction method and forms of hydration reaction to which they were subjected at the time.

Feng et al. (2023) mention that buildings are considered of utmost importance because they have to comply and become more sustainable, studies should be carried out in order to give a new use to certain components called waste, as a more suitable and sustainable alternative worldwide, new design alternatives to obtain quality concrete using new materials.

Given this, Simsek et al. (2022) state that it is one of the materials generated by the construction industry, since they produce large amounts of waste, of which are unwanted materials due to demolition, much of these materials may come from the crushed concrete elements of buildings and other structures, in another sense, if it comes to use the aggregate components of recycled concrete, the size of its particles becomes critical to the compressive strength of the development of new concrete.

According to Makul (2020), concrete is the second most consumed product on earth after water, likewise, it is a cement-bonded aggregate, the binder is largely Portland Cement, obtained by calcination of lime ($CaCO_3$) at high temperatures, on the other hand, the increasing demand for concrete structures comes to impose burdens on the environment and finite resources. In addition, Sun et al. (2020) indicate that concrete is one of the most useful and important building components in many countries, possessing many advantages, including high compressive strength, easy availability and ease of modelling, as well as being characterised by having 75% of aggregates of natural origin.

For Armany (2021) cement is a fine, soft, powdery type substance, which is mainly used to bind aggregates for concrete mixes, in addition cement acts as a hydraulic binder; it hardens during the addition of water, and is the key ingredient in concrete and mortar that will be used to build durable structures.

Dawoud et al. (2020) indicate that cement is considered one of the most important components worldwide, as it is the backbone for the progress of any country, being widely consumed building materials due to its excellent mechanical properties and durability. For Arivazhagan et al.(2020), concrete blocks are manufactured from sand, gravel and mixed with cement into blocks, largely referred to as concrete masonry units, being manufactured from pressed, cast or extruded aggregate concrete, they can be manufactured for virtually any architectural or structural function and their specifications for concrete masonry units are provided in various standards, including ASTM C55.

Likewise, according to Abasi et al. (2023), they mention that a large part of the masonry structures are recently making use of concrete blocks, given that their characteristics include low cost, high load capacity and energy efficiency, being one of the construction components widely used in the United States as they possess one of their properties of great magnitude, which is thermal resistance.

Concrete blocks are characterised by being one of the materials used in masonry, in addition to being low cost compared to traditional masonry units, on the other hand, within its classification can be found blocks with normal, medium or light weights, in addition to being characterised by being resistant, used for structural and decorative purposes (Zahra et al., 2021).

On the other hand, the regulation mentions that the masonry units for structural use will have to consider the minimum values stipulated in the regulation, as a first point it takes into account the blocks for non-load-bearing walls where the maximum warping will have to be 8mm, likewise the blocks for load-bearing walls will have to consider a maximum warping of 4mm, being important the shape of the final blocks that will form the structures with their corresponding characteristics (RNE, 2019).

Peng et al.(2022), indicate that another important test is the absorption of each masonry unit or concrete block, which is carried out in order to know the consistency of the blocks as well as the water absorption capacities that they may have up to the point of becoming saturated, the degrees of absorption will depend on the type of concrete that is formed, it is important that the units have lower saturation coefficients in order to avoid their porosity.

According to Cabané et al. (2022), they state that, in order to know the mechanical properties of concrete blocks, the compressive strength test of these blocks will have to be carried out, with the purpose of obtaining values of how much these units can resist crushing loads, determining their resistance limits and always considering compliance with the standard that is being taken into consideration for the respective test.

In addition, according to the masonry standard E.070, the compressive strength of the blocks to be studied will have to consider the minimum values stipulated by the standard given on the gross areas, where for the first point it mentions blocks for non load-bearing walls, which indicates that they must have minimum strengths of 20 kg/cm^2 , likewise for load-bearing wall blocks it will have to be considered for minimum strengths of 50 kg/cm^2 (RNE, 2019).

To assess the compressive strength of a concrete block, it must be tested in compression according to ASTM C140, and the load faces must be kept parallel so that the load can be evenly distributed over the entire area of the block.

The compression behaviour of concrete masonry blocks is one of the most important and necessary issues in the design and safety assessment of buildings because of the compressive forces that these structures are subjected to (Fakharian et al., 2023).

Regarding the bending strength, it is performed for the determination of the stiffness strengths of the specimens, where the specimens are placed on 2 supports, by means of the application of a centric load to the specimen in a set time, where the breaking loads are called bending strengths (Nikolenko et al., 2021).

On the other hand, the tensile strength, also known as the diametral test, is carried out by means of a universal machine where the tensile strength of the concrete is evaluated according to the ASTM standard, or by the NTP 399.621 in order to calculate the shear forces.

Tensile strength is one of the main parameters in concrete structural engineering and is a vital characteristic of concrete as structural elements are extremely vulnerable to cracking due to a variety of factors, including applied load, thus describing the ability of a component to resist tensile stress (Carrico et al., 2021).

Farrel et al (1967) in tests on soils with a very wide gradation, ranging from coarse sand to clay, found tensile strength values corresponding to approximately 10% of those of compressive strength (at/ac = 0.1), while Law (1987, quoted by Abu-Hejleh and Znidarcic, 1995) suggests that the ratio has a maximum value of 0.5.

Otazzi (2004) the ratio of tensile strength to compressive strength is within a range of 8 to 15% of the compressive strength.

Gerardo A. Rivera L. (2013) In his book Simple Engineer's Concrete, paragraph 6.6

founds correlations between flexural strength and compressive and tensile strengths, and concludes that the modulus of rupture shows values varying from 10% to 20% of the compressive strength. An approximate relationship to be used when flexural tests are not available is as follows: $\boldsymbol{MR = k(RC)^{112}}$

III.- METHODOLOGY

3.1. Type and design of research:

Type:

Villanueva (2022) mentions that the purpose of the applied study is to investigate and seek some kind of solution based on the theory to solve the problems by means of the objectives of the study.

Arias (2006) refers to the degree of study to deal with a phenomenon or object; the greater the depth, the better, and it can be exploratory, descriptive and explanatory.

Therefore, our research is based on the applied type, as we will use RCBCACR for the purpose of optimising the properties of concrete.

Design:

Ramírez and Calles (2021) indicate that within this design there is a group that will be exposed to pure experimental and another control group, where the first group will be manipulated causing certain changes, of which on many occasions what is sought is to generate an optimisation.

Palella and Martins (2006, p. 95), research design refers to a tactic that the researcher adopts to solve the problem, conflict or shortage posed in the study.

Because of this, this research will be of pure experimental design, due to the search for improvements of each property of the concrete by adding RCBCACR.

Focus:

Likewise, Sánchez (2019) mentions the quantitative study approach, through which data is obtained by means of data collection, and the objects of study become susceptible to being measured numerically.

Niño (2021) indicates that within the quantitative approach, numerical values are obtained, i.e. they will be expressed numerically, being very different from the qualitative approach, as its very name indicates, the qualities of the object under study.

Sampieri R. et al (2004), mentions that the quantitative approach is based on a methodical and logical scheme and attempts to formulate research questions and hypotheses for later demonstration.

Given this, a quantitative approach will be used because the tests that will be carried out will provide us with numerical data which will be measured, and these will be collected by means of the appropriate study instruments.

3.2. Variables and Operationalisation:

Independent Variable 1: Addition of sugar cane bagasse ash residues.

Conceptual definition: Sugar cane bagasse ash is defined as a material calcined at temperatures of 800°C to 1000 °C and has properties that favour high pozzolanic activity (Frías et al., 2007).

Operational definition: The measure needed for the respective concentrations of sugar cane bagasse ash to find the optimal percentage for use in building blocks.

Independent Variable 2: Addition of recycled concrete.

Conceptual definition: Recycled concrete or rubble dumps are defined as solid waste derived from the demolition of buildings, mostly from demolition and/or remodelling of dwellings, but also from changes in floors, paving or pavements (Ingenieros de Minas de Madrid, 2015).

Operational definition: The measure needed for the respective concentrations of recycled concrete to find the optimum percentage for use in block construction.

Dependent Variable: Concrete blocks

Conceptual definition: The blocks are generally made of cement, sand, gravel and water, the blocks have a hollow interior that allows the passage of steel bars and mortar filling, the dimensions vary from the most common which is 19x19x39 cm, for structural use, and other more aesthetic for masonry, with approximate measures of 19x9x39 (Tomás Franco, 2018).

Operational definition: Compressive, flexural and tensile strength tests will be carried out on the masonry blocks with different percentages of addition and the optimum will be known.

3.3. Population, sample and sampling:

Population

Arias (2006, p.81), defines population as a limited or unlimited set of elements of a common typology that will be extended by his research findings, being delimited by the problem and the study objectives.

Moguel (2005, p.85) defines limited population as when it is known how many elements the population has, he also defines unlimited population as part when the exact number of units of which the population is composed is not known.

The target population must be clearly and precisely delimited in the research problem and in the general objective of the study.

The population for the present study consisted of 180 blocks measuring 40x12x20cm per unit.

Sample

Castro (2003), the sample is classified into random and non-random. The random are all members of the population that have an equal chance of forming it in turn, they can be simple random sample, methodical random sample, stratified sample or cluster or regions. Non-random, the choice of members for the study is at the discretion of the researcher, meaning that not all members of the population assume an equal chance of forming it. The way to obtain this sample specimen is by purposive or opinion sampling and random or non-standard sampling.

Most authors agree that it is possible to take around 30% of the population and obtain a very representative sample (Ramirez, 1999, p. 91) and when the population is less than 50 individuals, the population is equal to the sample, according to Hernandez cited in (Castro, 2003, p. 69).

The sample will be equal to the population, i.e. 180 blocks of blocks; of which 45 blocks will be for determining the standard sample (5 standard samples for each test with ages of 7, 14 and 28 days) and 135 blocks will be for the different dosages to be optimised (5 samples per RCBCACR test of proportions 5%, 10%, and 15% with an age of 7, 14 and 28 days each test).

Sampling.

Hernández et al (2006) indicate that sampling attempts to explore the relationships between the distribution of the variable "y" in the population "z" and the distribution of the same variable in the sample under study.

Sampling as a group of operations carried out on the entire population universe to study the distribution of certain characteristics, based on the observation of a fraction

of the population considered (Tamayo and Tamayo, 2006, p. 176).
It is the way to find the possibility that each component forms a sample. Therefore, this procedure is performed by non-probabilistic random sampling and is used in populations b and c (Arias, 2006, p. 83). It is a technique used to select the blocks to be used in the tests.

Unit of analysis.

Hurtado (2000) the unit of study is referred to the context, being the possessor of the characteristics, event, variable or quality to be studied, a unit of study can be a person, object, group, geographical extension, institution, among others.
Omar at el. (2007) the relationship of unit of analysis and unit of observation is an extension of the notion of the data matrix proposed by Samaja at the VII Jornadas de Sociología at the University of Buenos Aires.
The unit will consist of the concrete block with the addition of RCBCACR.

3.4. Data collection techniques and instruments:

Data collection techniques

Direct observation will be used as a technique, because it involves the careful observation, recording and organisation of the effects that will be seen in the study by testing the mechanical properties of concrete blocks and recording the data obtained.

Data collection instruments

Material test forms will be used as a tool to help record the characteristics of the sample during the development of the study, thus organising and classifying all the data collected through the laboratory tests for our dosage, design and breakage tests.

Validity

This term refers to the degree to which an instrument becomes capable of measuring the variables allowing their respective evaluation that will be related to the purpose of the study, given this, each of the tests will be performed under optimal conditions with instruments and machines that relate to the required quality standards, and will be supported by a calibration certificate in order to ensure that the results that are achieved by the laboratory have not been altered in any way, since they will provide results with total authenticity and reliability when evaluated by experts in this subject.

Reliability

The purpose of this term is to determine the accuracy of the results obtained in various situations when using the respective instruments, and, accordingly, the reliability of the equipment and tests must be reliable to be carried out in the laboratory with the aid of classified equipment, in order to avoid obtaining erroneous results.

3.5. Procedures:

The Thesis Development Procedure Flowchart is presented (for more detail see MIRO *see figure 55 p. 142*).

Data collection:

First. - Recycled concrete will be collected from the accesses to the Estancia del Valle de Chiclayo housing estate.

Second. - The purchase of sugar cane bagasse ash from the Pucalá Agroindustry in Pomalca, which results from the combustion of sugar cane bagasse in the boiler at high temperatures, will be carried out.

Third. - The coarse sand will be acquired from the quarry San Nicolas - Pátapo - Ferreñafe - Lambayeque and the confetti from the quarry Centro - La Cría - Pátapo - Ferreñafe - Lambayeque.
Fourth. - Recycled concrete shall be cleaned to remove organic matter or other contaminants that affect the concrete and shall be crushed to sizes smaller than[1] Z¡".
Fifth. - We will proceed to process the physical and mechanical characteristics of the materials taking into account the Technical Standard E.070 Masonry and ASTM, such as: Granulometry of the aggregate according to NTP 400.012, NTP 400.037, NTP 339.185 moisture content, specific weight, absorption according to NTP 399.604 and 400.021 and unit weight according to NPT 400.017.
For the granulometric analysis, 500 g of sand sample is weighed, then it is stopped by the sieve, 3/8, 4, 8, 16, 30, 50, 100, 200. After shaking, each aggregate size is separated, each of these is weighed to measure the percentage that each one has with respect to the total weight.
Sixth. - The acquisition of QHUNA TYPE I cement and other materials to be used to help in the elaboration of the concrete blocks will be carried out.
Seventh. - The sugar cane bagasse ash residues shall be sieved to remove particles using sieve #16.
Eighth. - The mix design for the respective tests will be carried out, considering the standard concrete and the concrete with RCBCACR at 5%, 10% and 15%, using the ACI 211 method for the mix design, by means of a spreadsheet.
Ninth. - The dosage and elaboration of the concrete blocks for the respective tests will be carried out, considering the standard concrete and the concrete with RCBCACR at 5%, 10% and 15%, using the ACI 211 method for the design of mixtures.
Tenth. - The concrete blocks (sample) will be made by hand using the dosage indicated for each sample, having as inputs yellow sand, confetti from[1] Λ to #16, cement, recycled concrete, cement, water.
Eleventh. - The samples shall be broken in a period of 7, 14 and 28 days, for the compressive, flexural and tensile strength tests, according to NTP 399.604 and 399.6013, according to NTP 339.079 and according to NTP 339.084.
Twelfth. - Finally, the data obtained in the tests will be compiled and recorded in Excel, and a comparison of these data will be made, in order to draw conclusions and recommendations in the final report.

3.6. Method of data analysis:

In the present investigation, the data will be processed and analysed using the instruments already standardised for each of the tests, the data will be collected and processed through Excel spreadsheets, where the values will be represented by means of tables and graphs, evidencing each of the results, which will help to verify whether the results benefit or solve the problem posed in the study, in order to be analysed in more detail and better visualisation.

3.7. Ethical aspects:

The author of the study will comply with the ethical principles established in the Code of Ethics of the Universidad César Vallejo, which is based on integrity in each scientific study and management work, intellectual honesty in each aspect covered by the study, objectivity and impartiality which is based on how to relate to work and professionally; truthfulness, fairness and responsibility to execute and disseminate each result achieved through the study, transparency which is based on acting without any conflict of interest, to declare and manage these conflicts whether economic or other issues in reference, autonomy where each participant in the study will be free to participate or withdraw from it at the time they believe fit for it, care for the environment where each researcher must provide the assurance to care for and respect the natural environment without facts that affect it, human integrity by putting the person before scientific interest, social status, gender or other parameters in reference, equity, including equal treatment for each participant in the study, intellectual respect for the citation rights of each national and international author according to ISO 690, privacy will be based on safeguarding the information collected in a secure manner, finally independence, where this refers to the fact that the study will be autonomous, as well as guaranteeing the absence of a high plagiarism rate, which will be checked with the Turnitin anti-plagiarism tool.

IV.- RESULTS

Figure 1

Grain-size curve for coarse sand.

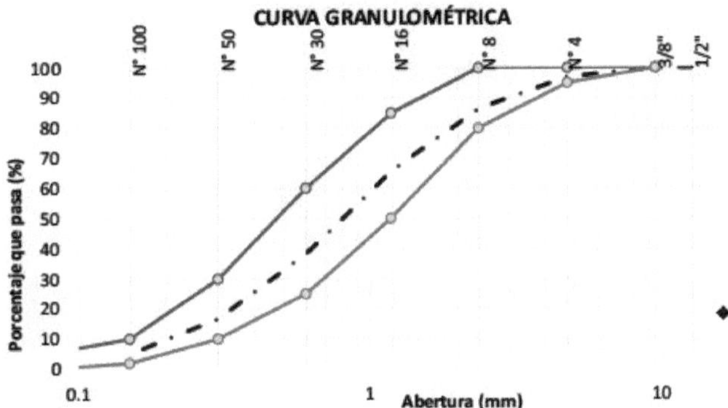

Note. This figure shows the behaviour of the sand and is between the mesh N°3/8 and mesh N°100, being suitable for use.

Figure 2

Confitillo Granulometric Curve.

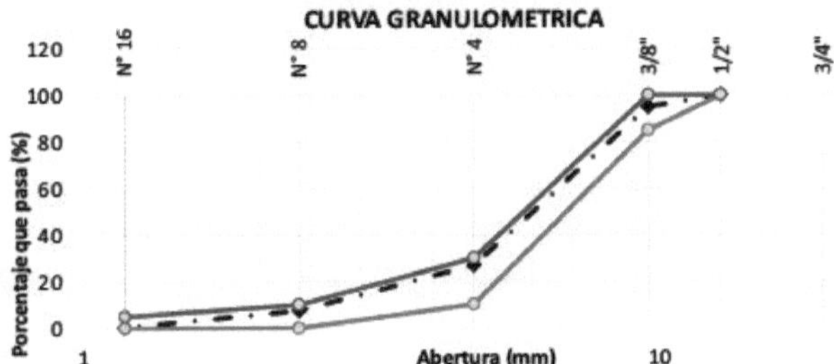

Note: This figure shows the behaviour of the candyfloss and is between mesh N°1/2 and mesh N°16, being suitable for use.

Figure 3

Granulometric Curve Recycled Concrete.

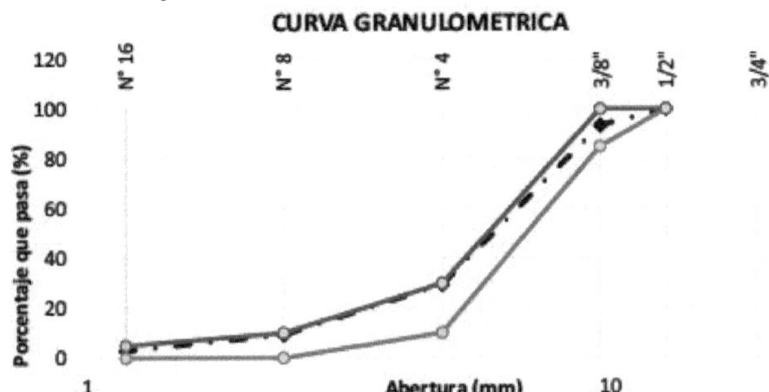

Note: This figure shows the behaviour of recycled concrete and is between No. 1/2

mesh and No. 16 mesh and is suitable for use.

Figure 4

Moisture content of aggregates.

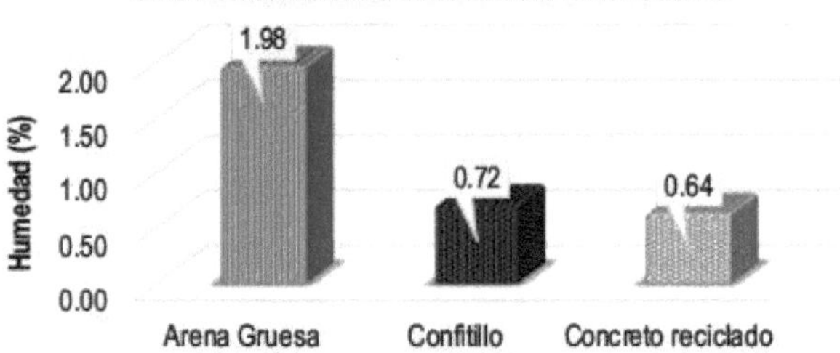

Note: This figure shows that the moisture content of recycled concrete is similar to that of ready-mixed concrete, so it can behave like ready-mixed concrete.

Figure 5

Specific gravity of aggregates.

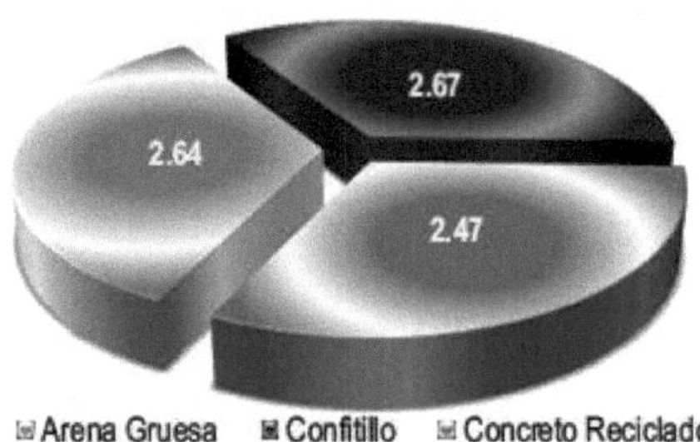

Note: This figure shows that the specific gravity of recycled concrete is similar to that of ready-mixed concrete, so it can behave like ready-mixed concrete.

Figure 6

Percentage Absorption of Aggregates.

Note: This figure shows that the absorption rate of recycled concrete is similar to that of ready-mixed concrete, so it can behave like ready-mixed concrete.

Figure 7

Unit weight of aggregates.

Note: This figure shows that the unit weight of loose and compacted coarse sand for the mix design.

Figure 8

Unit weight of Confitillo

Note: This figure shows that the unit weight of the loose and compacted ream, for the mix design.

Figure 9

Unit Weight of Recycled Concrete.

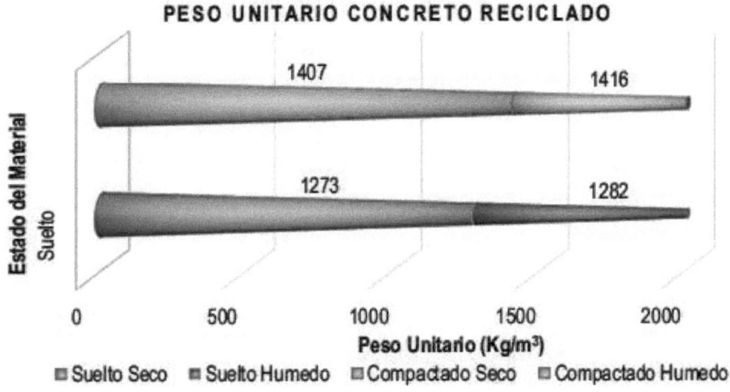

Note: This figure shows that the unit weight of loose and compacted recycled concrete behaves similarly to that of ready-mixed concrete, so it can be replaced in the mix design.

ACI 211 Mixture Design: With addition of 0% RCBCACR

Resilient design				50 Kg/cm 2	
I.) Coarse aggregate data				Confitillo	
01.-	Nominal maximum size			3/8"	in.
02.-	Specific dry mass weight			2630.32	Kg/m^3
03.-	Dry compacted unit weight			1415.17	Kg/m^3
04.-	Dry loose unit weight			1288.92	Kg/m^3
05.-	Moisture content			0.72	%
06.-	Absorption content			0.63	%
II.) Fine aggregate data				Coarse Sand	
07.-	Specific dry mass weight			2519.27	Kg/m^3
08.-	Dry loose unit weight			1422.09	Kg/m^3
09.-	Moisture content			1.98	%
10.-	Absorption content			1.83	%
11.-	Modulus of fineness (dimensionless)			2.91	
III.) Mixing and other data					
12.-	Water-cement ratio			$R^{a/c}$ 0.87	
13.-	Settlement			1-2	In.
14.-	Unit volume of water		Drinking water in the area	207.00	L/m^3
15.-	Trapped air content			3.00	%
16.-	Coarse aggregate volume			0.45	m^3
17.-	Specific gravity of cement		QHUNATYPE 1	3850.00	Kg/m^3

IV.) Calculation of absolute volumes, correction for humidity and water input					
a.- **C e m m e n t s**	237.90	0.062			
b.- **A g u a**	207.00	0.207			
c.- **A i r e**	3.00	0.030		Moisture	Effective Water 2.3
d.- **A r e n a**	1156.34	0.459	1612.56	correction 1644.43	0.17
e.- **C o n f i t i 11 o f**	635.39	0.242	179.17	180.47	
V.) Final design result (wet)		0.541	VI.) Test run		0.024495 m^3
CEMENT	237.90 kg/m^3		5.83 kg	p /cement (in bags)	5.60
WATER	204.52 kg/m^3		5.01 kg	R a/c design	0.87
ARENA	1644.43 kg/m^3		40.29 kg	on-site a/c	0.86
CONFITILLO	180.47 kg/m^3		4.43 kg		
	2267		55.60 kg		

VII). Volume dosage (materials with		**natural moisture)**			
	Cement	Sand	Confitillo	Water	
In 1 ft bag3 P	1.0	6.91	0.76	36.54	Lts/ft^3
In 1 ft bag3 V	**1.0**	**7.29**	**0.88**	**36.54**	$\mathbf{Lts/ft^3}$

Note: This table shows the mixing design and dosage without addition of RCBCACR.

ACI 211 Mixture Design: With addition of 5% RCBCACR

I.) Coarse aggregate data		**Confittilo**	
01.-	Nominal maximum size	3/8"	in.
02.-	Specific dry mass weight	2630.32	Kg/m3
03.-	Dry compacted unit weight	1415.17	Kg/m3
04.-	Dry loose unit weight	1288.92	Kg/m3

05.-	Moisture content		0.72	%
06.-	Absorption content		0.63	%
II.) Fine aggregate data			**Coarse Sand**	
07.-	Specific dry mass weight		2519.27	Kg/m^3
08.-	Dry loose unit weight		1422.09	Kg/m^3
09.-	Moisture content		1.98	%
10.-	Absorption content		1.83	%
11.-	Modulus of fineness (dimensionless)		2.91	
III.) Mixing and other data				
12.-	Water-cement ratio		$R^{a/c}$ 0.87	
13.-	Settlement		1-2	In.
14.-	Unit volume of water	Drinking water in the area	207.00	L/m^3
15.-	Trapped air content		3.00	%
16.-	Coarse aggregate volume		0.45	m^3
17.-	Specific gravity of cement	QHUNATYPE 1	3850.00	Kg/m^3

IV.) Calculation of absolute volumes, correction for humidity and water input						
a.- C e m m e n t s	237.90	0.062				
b.- Ag u a	207.00	0.207				
c.- A i r e	3.0	0.030		Correction by humidity	Water Effective	
d.- A r e n a	1156.34	0.459	1612.56	1644.43	2.3	
e.- C o n f i t i 11 o f	635.39	0.242	179.17	180.47	0.17	
V.) Final design result (wet)		0.541	VI.) Probe Test Round		0.024495	m^3
CEMENT	226.01 kg/m^3		5.53 kg		p/cement (in bags)	5.32
ASH	11.90 kg/m^3		0.30 kg			
WATER	204.52 kg/m^3		5.01 kg		R a/c design	0.87
ARENA	1644.43 kg/m^3		40.29 kg		p a/c de obra	0.86
WITH PITI LLO	171.45 kg/m^3		4.20 kg			
RECYCLED CONCRETE	9.02 kg/m^3		0.23 kg			

VII). Volume dosage (materials with natural moisture)

	Cement	Arena	Confitil lo	Ash	Concrete	Water	
In 1 ft bag3 P	1.0	7.28	0.76	0.05	R. 0.04	38.46	Lts/ft^3
In 1 ft bag3 V	**1.0**	**7.68**	**0.88**	**0.06**	**0.05**	**38.46**	**Lts/ft^3**

Note: This table shows the mixing design and dosage with addition of 5% RCBCACR.

Table 3

ACI 211 Mixture Design: With addition of 10% RCBCACR

I.) Coarse aggregate data	**Confittilo**		
01.-	Nominal maximum size	3/8"	in.
02.-	Specific dry mass weight	2630.32	Kg/m^3
03.-	Dry compacted unit weight	1415.17	Kg/m^3
04.-	Dry loose unit weight	1288.92	Kg/m^3
05.-	Moisture content	0.72	%
06.-	Absorption content	0.63	%

II.) Fine aggregate data			Coarse Sand	
07.-	Specific dry mass weight		2519.27	Kg/m^3
08.-	Dry loose unit weight		1422.09	Kg/m^3
09.-	Moisture content		1.98	%
10.-	Absorption content		1.83	%
11.-	Modulus of fineness (dimensionless)		2.91	
III.) Mixing and other data				
12.-	Water-cement ratio		$R^{a/c}$0.87	
13.-	Settlement		1-2	In.
14.-	Unit volume of water	: Drinking water from the zone	207	L/m^3
15.-	Trapped air content		3	%
16.-	Coarse aggregate volume		0.45	m^3
17.-	Specific gravity of cement	QHUNATYPE 1	3850	Kg/m^3

IV.) Calculation of absolute volumes, correction for humidity and water input

a.- Cemments	237.90	0.062				
b.- Agua	207.00	0.207			Effective Water	
c.- Aire	3.0	0.030		Moisture correction		
d.-Sand	1156.34	0.459	1612.56	1644.43	2.3	
e.- Confiti11of	635.39	0.242	179.17	180.47	0.17	
V.) Final design result (wet)		0.541	**VI.) Probe Test Round**		**0.024495**	m^3
CEMENT	$214.11 kg/m^3$		5.23 kg		p/cement (in bags)	5.04
ASH	$23.79 kg/m^3$		0.60 kg		R a/c of design p a/c of construction site	0.87
WATER	$204.52 kg/m^3$		5.01 kg			0.86
ARENA	1644.43		40.29 kg			
WITH PITI LLO	kg/m^3		3.97 kg			
RECYCLED CONCRETE	$162.42 kg/m^3$		0.46 kg			
	$18.05 kg/m^3$					

VII). Volume dosage (materials with natural moisture)

	Cement	Sand	Confidilo	Ash	Concrete R.	Water	
In 1 ft bag3 P	1.0	7.68	0.76	0.11	0.08	40.60	Lts/ft^3
In 1 ft bag3 V	**1.0**	**8.10**	**0.88**	**0.13**	**0.10**	**40.60**	Lts/ft^3

Note: This table shows the mixing design and dosage with addition of 10% RCBCACR.

ACI 211 Mixture Design: With addition of 15% RCBCACR

I.) Coarse aggregate data Confidilo			
01.-	Nominal maximum size	3/8"	in.
02.-	Specific dry mass weight	2630.32	Kg/m^3
03.-	Dry compacted unit weight	1415.17	Kg/m^3
04.-	Dry loose unit weight	1288.92	Kg/m^3
05.-	Moisture content	0.72	%
06.-	Absorption content	0.63	%
II.) Fine aggregate data		**Coarse Sand**	
07.-	Specific dry mass weight	2519.27	Kg/m^3
08.-	Dry loose unit weight	1422.09	Kg/m^3
09.-	Moisture content	1.98	%

10.-	Absorption content		1.83	%
11.-	Modulus of fineness (dimensionless)		2.91	
III.) Mixing and other data				
12.-	Water-cement ratio		$R^{a/C}$0.87	
13.-	Settlement			1-2 In.
14.-	Unit volume of water	Drinking water in the area		207 L/m^3
15.-	Trapped air content			3%
16.-	Coarse aggregate volume		0.44898	m^3
17.-	Specific gravity of cement	QHUNA TYPE I		3850 Kg/m^3

IV.) Calculation of absolute volumes, correction for humidity and water input

a.- C e m m e n t s	237.90	0.062				
b.- A g u a	207.00	0.207	Moisture	correction Water		Effective
c.- A i r e	3.0	0.030				
d.- A r e n a	1156.34	0. 4591612.	561644. 432.3			
e.- C o n f i t i 11 o f	635.39	0.242 179.17		180.47	0.17	
	2240	1.000			2.48	

V.) Final design result (wet)		0.541 **VI.) Probe Test Round**		0.024495 m^3
CEMENT	202.22 kg/m^3	4.93 kg	p/cement (in bags)	4.76
ASH	35.69 kg/m^3	0.90 kg		
WATER	204.52 kg/m^3	5.01 kg	p a/c of design	0.87
ARENA	1644.43 kg/m^3	40.29 kg	on-site a/c	0.86
CONFITILLO	153.40 kg/m^3	3.74 kg		
RECYCLED CONCRETE	27.07 kg/m^3	0.69 kg		

VII). Volume dosage (materials with natural moisture)

	Cement	Sand	Confitillo	Ceniza	Concrete R.	Water	
In 1 ft bag3 P	1.0	8.13	0.760	.18	0.13	42.98	Lts/ft^3
In 1 ft bag3 V	**1.0**	**8.58**	**0.880**	**.21**	**0.16**	**42.98**	**Lts/ft^3**

Note: This table shows the mixing design and dosage with the addition of 15% RCBCACR.

Aggregate dosage for a batch of 5 samples.

% ADDITION	**% ADDITION OF SUGAR CANE BAGASSE ASH RESIDUES AND RECYCLED CONCRETE TO BE USED IN THE 5-SAMPLE BATCH DESIGN**					
	CEMENT (KG)	ASH (KG)	WATER (LTS)	SAND (KG)	CANDY (KG)	RECYCLED CONCRETE(KG)
0%	5.83	0.00	5.01	40.29	4.43	0.00
5%	5.53	0.30	5.01	40.29	4.20	0.23
10%	**5.23**	**0.60**	**5.01**	**40.29**	**3.97**	**0.46**
15%	4.93	0.90	5.01	40.29	3.74	0.69

Note: This table shows the dosage with addition of RCBCACR at 0%, 5%, 10% and 15%, for a batch of 5 samples.

Table 6

Percentage Addition of RCBCACR.

% ADDITION	**% ADDITION OF BAGASSE ASH RESIDUES FROM SUGAR CANE BAGASSE SUGAR AND RECYCLED CONCRETE TO BE USED ON SITE FOR A 42.5 KG BAG OF CEMENT**

	CEMENT (KG)	ASH (KG)	WATER (LTS)	SAND (KG)	CANDY (KG)	RECYCLED CONCRETE(KG)
0%	42.50	40.36	38.24	36.10	42.50	40.36
5%	0.00	2.13	4.27	6.40	0.00	2.13
10%	**36.54**	**36.54**	**36.54**	**36.54**	**36.54**	**36.54**
15%	293.77	293.77	293.77	293.77	293.77	293.77

Note: This table shows a dosage for a 42.50 kg bag of cement for each dosage of 0%, 5%, 10% and 15% RCBCACR addition.

Compressive strength of each sample with 0% addition of RCBCARC

Nºof Witnesses	Description	Date of breakage Moulding	Breakage	Age (days)	Length (cm)	Width (cm)	High (cm)	Thickness (cm)	Area (cm 2)	Charge Kg	Obtained resistance F'b (kg/cm)²	Average F'b strength (kg/cm)²	Strength F'b Design (kg/cm)²	%
1.00	C - 0% Ash and R.C.	29/05/2023	5/06/2023	7	39.50	12.00	19.80	2.50	267.50	11.790.00	44.07			
2.00	C - 0% Ash and R.C.	29/05/2023	5/06/2023	7	39.40	11.90	19.80	2.60	274.56	10.490.00	38.21			
3.00	C - 0% Ash and R.C.	29/05/2023	5/06/2023	7	39.40	11.80	19.90	2.40	256.32	11.410.00	44.51			
4.00	C - 0% Ash and R.C.	29/05/2023	5/06/2023	7	39.60	12.00	19.80	2.50	268.00	10.920.00	40.75			
5.00	C - 0% Ash and R.C.	29/05/2023	5/06/2023	7	39.50	12.00	19.80	2.30	249.78	11.600.00	46.44	39.51	50.00	79 %
6.00	C - 0% Ash and R.C.	29/05/2023	12/06/2023	14	39.30	11.90	19.90	2.60	274.04	12.370.00	45.14			
7.00	C - 0% Ash and R.C.	29/05/2023	12/06/2023	14	39.40	11.90	19.70	2.30	248.40	13.510.00	54.39			
8.00	C - 0% Ash and R.C.	29/05/2023	12/06/2023	14	39.40	12.00	19.80	2.50	267.00	13.390.00	50.15			
9.00	C - 0% Ash and R.C.	29/05/2023	12/06/2023	14	39.50	11.90	19.80	2.40	257.76	12.250.00	47.52			
10.00	C - 0% Ash and R.C.	29/05/2023	12/06/2023	14	39.50	12.00	19.90	2.50	267.50	12.970.00	48.49	45.69	50.00	91 %
11.00	C - 0% Ash and R.C.	29/05/2023	26/06/2023	28	39.40	11.80	19.80	2.60	273.52	14.450.00	52.83			
12.00	C - 0% Ash and R.C.	29/05/2023	26/06/2023	28	39.40	12.00	19.90	2.50	267.00	13.770.00	51.57			
13.00	C - 0%	29/05/2	26/06/2	28	39.50	11.9	19.	2.40	256.	13.530.	52.69			

	Ash and R.C.	023	023			0	80		80	00				
14.00	C - 0% Ash and R.C.	29/05/2023	26/06/2023	28	39.50	11.80	19.70	2.30	248.86	14.370.00	57.74			
15.00	C - 0% Ash and R.C.	29/05/2023	26/06/2023	28	39.40	11.90	19.80	2.50	230.16	14.110.00	61.31	51.08	50.00	102 %

Note: This table shows the addition of 0% RCBCACR, having a compressive strength of 51.08 kg/cm^2 at 28 days.

Compressive strength of each sample with 5% addition of RCBCARC

№ of Witnesses	Description	Date of breakage		Age (days)	Length (cm)	Width (cm)	High (cm)	Thickness (cm)	Area (cm2)	Charge Kg	Obtained resistance F'b (kg/cm)^2	Resistance . Average F'b (kg/cm)^2	Strength F'b Design (kg/cm)^2	%
		Moulding	Breakage											
1.00	C - 5% Ash and R.C.	30/05/2023	6/06/2023	7	39.40	12.00	19.90	2.40	258.24	10.250.00	39.69			
2.00	C - 5% Ash and R.C.	30/05/2023	6/06/2023	7	39.40	12.00	19.90	2.30	249.32	10.800.00	43.32			
3.00	C - 5% Ash and R.C.	30/05/2023	6/06/2023	7	39.50	11.90	19.80	2.50	266.50	10.070.00	37.79			
4.00	C - 5% Ash and R.C.	30/05/2023	6/06/2023	7	39.40	12.00	19.80	2.50	267.00	10.750.00	40.26			
5.00	C - 5% Ash and R.C.	30/05/2023	6/06/2023	7	39.50	12.00	19.90	2.50	267.50	11.600.00	43.36	38.46	50.00	77 %
t 6.00	C - 5% Ash and R.C.	30/05/2023	13/06/2023	14	39.60	11.90	19.80	2.60	275.60	10.410.00	37.77			
7.00	C - 5% Ash and R.C.	30/05/2023	13/06/2023	14	39.40	11.90	19.80	2.40	257.28	11.430.00	44.43			
8.00	C - 5% Ash and R.C.	30/05/2023	13/06/2023	14	39.60	12.00	19.80	2.50	268.00	11.300.00	42.16			
9.00	C - 5% Ash and R.C.	30/05/2023	13/06/2023	14	39.50	11.90	19.70	2.50	266.50	12.650.00	47.47			
10.00	C - 5% Ash and R.C.	30/05/2023	13/06/2023	14	39.50	11.80	19.90	2.40	256.80	10.520.00	40.97	38.91	50.00	78 %
11.00	C - 5% Ash and R.C.	30/05/2023	27/06/2023	28	39.40	12.00	19.90	2.50	267.00	11.910.00	44.61			
12.00	C - 5% Ash and R.C.	30/05/2023	27/06/2023	28	39.40	11.90	19.80	2.60	274.56	13.130.00	47.82			
13.00	C - 5%	30/05/2	27/06/2	28	39.50	12.0	19.8	2.50	267.	11.600.	43.36			

	Ash and R.C.	023	023			0	0		50	00				
14.00	C - 5% Ash and R.C.	30/05/2023	27/06/2023	28	39.50	12.00	19.90	2.40	258.72	11.070.00	42.79			
15.00	C - 5% Ash and R.C.	30/05/2023	27/06/2023	28	39.40	12.00	19.90	2.60	275.60	12.160.00	44.12	42.58	50.00	85 %

Note: This table shows the addition of 5% RCBCACR, having a compressive strength of 42.58 kg/cm^2 at 28 days.

Compressive strength of concrete blocks with 0% concrete admixture RCBCACR.

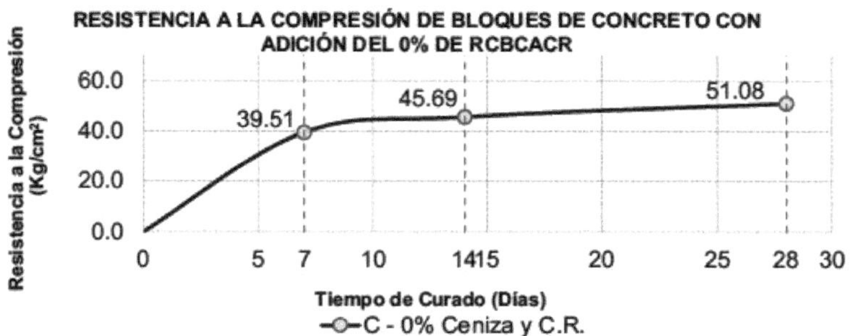

Note: This figure shows the addition of 0% RCBCACR, having a compressive strength of 51.08 kg/cm^2 at 28 days, being our standard sample.

Figure 11

Compressive strength of concrete blocks with 5% concrete admixture RCBCACR.

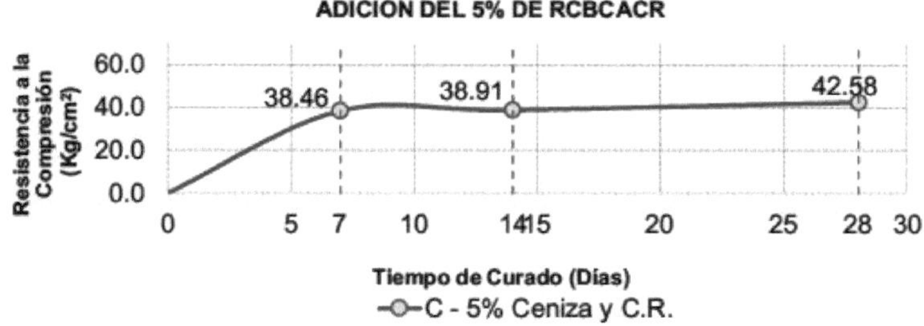

Note: This figure shows the addition of 5% RCBCACR, having a compressive strength of 42.58 kg/cm^2 at 28 days, below the standard sample.

Compressive strength of each sample with 10% addition of RCBCARC

№ of Witnesses	Description	Date of breakage		Age (days)	Length (cm)	Width (cm)	High (cm)	Thickness (cm)	Area (cm2)	Charge Kg	Resistance obtained F'b (kg/cm)^2	Average F'b strength (kg/cm)^2	Strength F'b Design (kg/cm)^2	%
		Moulding	Breakage											

1.00	C -10% Ash and R.C.	30/05/2023	6/06/2023	7	39.50	11.90	19.80	2.50	262.15	11.840.00	45.16			
2.00	C-10% ash and C.R.	30/05/2023	6/06/2023	7	39.50	11.90	19.80	2.60	275.08	10.800.00	39.26			
3.00	C-10% ash and R.C.	30/05/2023	6/06/2023	7	39.50	11.90	19.90	2.60	275.08	12.480.00	45.37			
4.00	C-10% ash and R.C.	30/05/2023	6/06/2023	7	39.40	12.00	19.90	2.50	267.00	12.190.00	45.66			
5.00	C -10% ash and C.R.	30/05/2023	6/06/2023	7	39.50	12.00	19.80	2.50	267.50	11.500.00	42.99	41.00	50.00	82 %
6.00	C -10% ash and C.R.	30/05/2023	13/06/2023	14	39.40	12.00	19.70	2.50	267.00	13.080.00	48.99			
è 7.00	C -10% ash and C.R.	30/05/2023	13/06/2023	14	39.50	12.00	19.90	2.30	249.78	13.360.00	53.49			
8.00	C -10% ash and C.R.	30/05/2023	13/06/2023	14	39.40	11.90	19.80	2.50	266.00	12.400.00	46.62			
9.00	C -10% ash and C.R.	30/05/2023	13/06/2023	14	39.60	11.90	19.80	2.40	258.24	13.520.00	52.35			
10.00	C -10% ash and C.R.	30/05/2023	13/06/2023	14	39.50	12.00	19.80	2.40	258.72	12.980.00	50.17	47.60	50.00	95 %
11.00	C-10% ash and C.R.	30/05/2023	27/06/2023	28	39.40	11.80	19.90	2.50	265.00	16.540.00	62.42			
12.00	C -10% ash and C.R.	30/05/2023	27/06/2023	28	39.50	12.00	19.90	2.40	258.72	15.730.00	60.80			
13.00	C -10% ash and C.R.	30/05/2023	27/06/2023	28	39.40	12.00	19.80	2.60	275.60	16.490.00	59.83			
14.00	C -10% ash and C.R.	30/05/2023	27/06/2023	28	39.60	11.90	19.80	2.50	267.00	14.440.00	54.08			
15.00	C -10% ash and C.R.	30/05/2023	27/06/2023	28	39.60	11.90	19.80	2.50	267.00	14.140.00	52.96	53.79	50.00	108 %

Note: This table shows the addition of 10% RCBCACR, having a compressive strength of 53.79 kg/cm^2 at 28 days.

Compressive strength of each sample with 15% addition of RCBCARC

№ of Witnesses	**Description**	**Date of breakage Moulding**	**Breakage**	**Age (days)**	**Length (cm)**	**Width (cm)**	**High (cm)**	**Thickness (cm)**	**Area (cm2)**	**Charge Kg**	**Resistance obtained F'b (kg/cm)^2**	**Average F'b strength (kg/cm)^2**	**Strength F'b Design (kg/cm)^2**	**%**
1.00	C-15%	30/05/2	6/06/20	7	39.50	12.0	19.	2.40	258.	11.300.	43.68			

	Ash and R.C.	023	23			0	90		72	00				
2.00	C-15% Ash and R.C.	30/05/2023	6/06/2023	7	39.40	11.80	19.80	2.50	265.00	10.450.00	39.43			
3.00	C-15% Ash and R.C.	30/05/2023	6/06/2023	7	39.50	11.80	19.80	2.60	274.04	10.460.00	38.17			
4.00	C-15% Ash and R.C.	30/05/2023	6/06/2023	7	39.60	11.90	19.80	2.50	256.47	10.620.00	41.41			
5.00	C-15% Ash and R.C.	30/05/2023	6/06/2023	7	39.50	11.90	19.90	2.50	262.15	11.600.00	44.25	38.76	50.00	78%
6.00	C-15% Ash and R.C.	30/05/2023	13/06/2023	14	39.50	11.80	19.80	2.50	263.77	10.520.00	39.88			
* 7.00	C-15% Ash and R.C.	30/05/2023	13/06/2023	14	39.40	12.00	19.80	2.50	267.00	11.820.00	44.27			
8.00	C-15% Ash and R.C.	30/05/2023	13/06/2023	14	39.40	12.00	19.90	2.60	275.60	11.430.00	41.47			
9.00	C-15% Ash and R.C.	30/05/2023	13/06/2023	14	39.50	11.90	19.90	2.60	275.08	12.480.00	45.37			
10.00	C-15% Ash and R.C.	30/05/2023	13/06/2023	14	39.60	11.90	19.80	2.50	267.00	12.020.00	45.02	40.80	50.00	82%
11.00	C-15% Ash and R.C.	30/05/2023	27/06/2023	28	39.50	11.80	19.80	2.60	274.04	12.370.00	45.14			
12.00	C-15% Ash and R.C.	30/05/2023	27/06/2023	28	39.40	12.00	19.90	2.60	275.60	13.390.00	48.58			
13.00	C-15% Ash and R.C.	30/05/2023	27/06/2023	28	39.50	12.00	19.80	2.50	267.50	13.530.00	50.58			
14.00	C-15% Ash and R.C.	30/05/2023	27/06/2023	28	39.50	11.90	19.80	2.60	275.08	13.660.00	49.66			
15.00	C-15% Ash and R.C.	30/05/2023	27/06/2023	28	39.40	11.90	19.90	2.60	274.56	13.080.00	47.64	46.23	50.00	92%

Note: This table shows the addition of 15% RCBCACR, having a compressive strength of 46.23 kg/cm^2 at 28 days.

Compressive strength of concrete blocks with 10% concrete admixture RCBCACR.

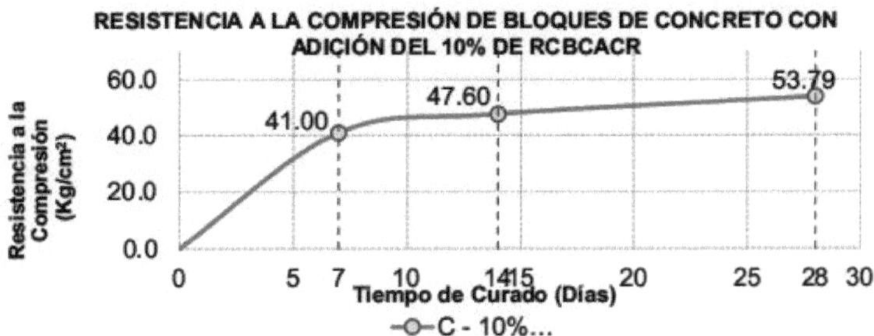

Note: This figure shows the addition of 10% RCBCACR, having a compressive strength of 53.79 kg/cm^2 at 28 days, above the standard sample.

Figure 13

Compressive strength of concrete blocks with 15% addition of RCBCACR.

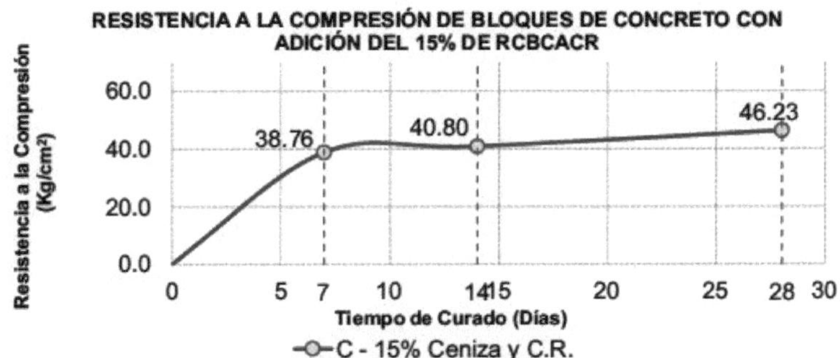

Note: This figure shows the addition of 15% RCBCACR, having a compressive strength of 46.23 kg/cm^2 at 28 days, below the standard sample.

Consolidated compressive strength of samples with 0%, 5%, 10%, 15%) addition of RCBCARC

Description	Date of breakage		Age (days)	Average F'b strength (kg/cm)^2	Strength F'b Design (kg/cm)^2	(%)
	Moulding	Breakage				
C - 0% Ash and R.C.	29/05/2023	5/06/2023	7.00	39.51	-	-
C - 0% Ash and R.C.	29/05/2023	12/06/2023	14.00	45.69	-	-
C - 0% Ash and R.C.	29/05/2023	26/06/2023	28.00	51.08	-	-
C - 5% Ash and R.C.	30/06/2023	7/07/2023	7	38.46	51.08	75%
C - 5% Ash and R.C.	30/06/2023	14/07/2023	14	38.91	51.08	76%
C - 5% Ash and R.C.	30/06/2023	28/07/2023	28	42.58	51.08	83%
C -10% ash and C.R.	30/05/2023	6/06/2023	7	41.00	51.08	80%
C -10% ash and C.R.	30/05/2023	13/06/2023	14	47.60	51.08	93%
C-10% ash and C.R.	**30/05/2023**	**27/06/2023**	**28**	**53.79**	**51.08**	**105%**
C -15% Ash and R.C.	30/05/2023	6/06/2023	7	38.76	51.08	76%
C -15% Ash and R.C.	30/05/2023	13/06/2023	14	40.80	51.08	80%
C -15% Ash and R.C.	30/05/2023	27/06/2023	28	46.23	51.08	90%

Note: This table shows the final results of the compressive strength with RCBCACR addition at 0%, 5%, 10% and 15%, for a batch of 5 samples, with the 10% dose being the best strength with 53.79 kg/cm .^{2}

Compressive strength of concrete blocks with addition of RCBCACR.

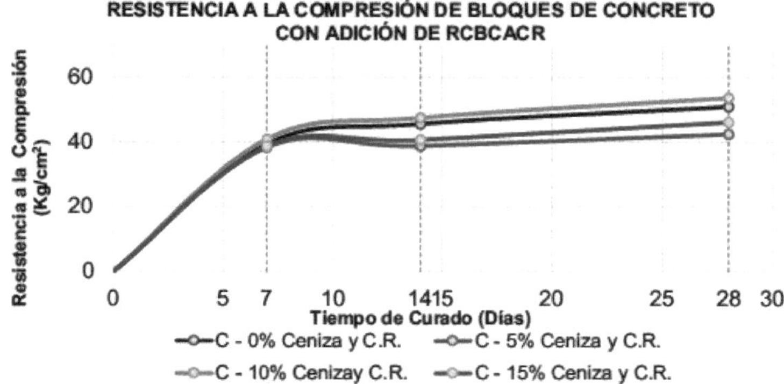

Note: This figure shows the addition of 10% RCBCACR, outperforms the other samples in strength, having a compressive strength of 53.79 kg/cm^2 at 28 days, even higher than the standard sample.

Table 12

Summary of compressive strength of concrete blocks with addition of RCBCACR at 0%, 5%, 10% and 15%.

	COMPRESSIVE STRENGTH (Kg/cm)^2		
Sample	**7 days**	**14 days**	**28 days**
0%	39.51	45.69	51.08
5%	38.46	38.91	42.58
10%	41.00	47.60	53.79
15%	38.76	40.80	46.23

Note: This table shows a summary of the compressive strength of the samples where the addition of 10% RCBCACR, exceeds in strength with 53.79 kg/cm^2 at 28 days.

Compressive strength of concrete blocks with addition of RCBCACR.

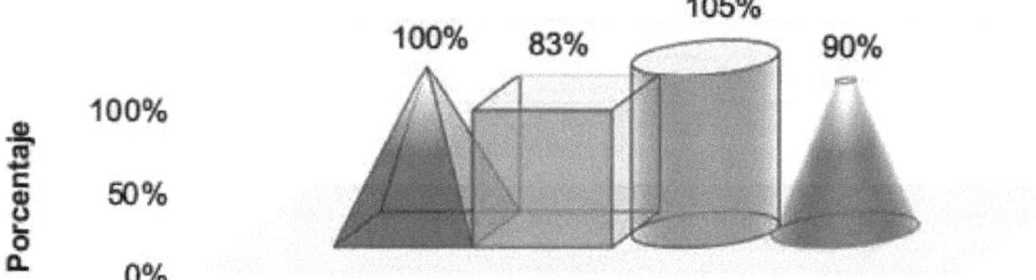

Note: This figure presents the results in average percentages of the concrete blocks with the addition of RCBCACR in 5%, 10% and 15%, of the 3 groups, the addition of 10% RCBCACR at 28 days of age is the one that presents the highest compressive strength with respect to the other groups, even surpassing the standard sample, concluding with the general and specific hypothesis if they comply, since the compressive strength improves significantly with the addition of RCBCACR, in addition they are within the range according to the E.070 Standard. which indicates that the minimum resistance for load-bearing blocks should be 50kg/cm^2 . It is concluded that the addition of 10% RCBCACR is the most optimal and can work as NP and P block, and that, if the addition of 10% is decreased or increased, the compressive strength decreases.

Table 13

Flexural strength of each specimen with 0% addition of RCBCARC

Description	Date Moulding	Date Breakage	Age (days)	Length (cm)	Width (cm)	Height (cm)	Free span between supports (cm)	Charge Kg	Mr (Kg/cm)^2	Average Mr (Kg/cm)^2
F - 0% Ash and C.R.	1/06/2023	8/06/2023	7	39.50	12.00	19.80	34.00	300	3.25	
F - 0% Ash and C.R.	1/06/2023	8/06/2023	7	39.40	12.00	19.90	34.00	210	2.25	
F - 0% Ash and C.R.	1/06/2023	8/06/2023	7	39.40	11.90	19.80	34.00	220	2.41	
F - 0% Ash and C.R.	1/06/2023	8/06/2023	7	39.50	11.90	19.70	34.00	320	3.53	
F - 0% Ash and C.R.	1/06/2023	8/06/2023	7	39.40	12.00	19.80	34.00	270	2.93	**2.87**
F - 0% Ash and C.R.	1/06/2023	15/06/2023	14	39.60	11.80	19.80	34.00	420	4.63	
F - 0% Ash and C.R.	1/06/2023	15/06/2023	14	39.40	11.90	19.90	34.00	360	3.90	
F - 0% Ash and C.R.	1/06/2023	15/06/2023	14	39.40	11.90	19.90	34.00	390	4.22	
F - 0% Ash and C.R.	1/06/2023	15/06/2023	14	39.50	11.80	19.80	34.00	410	4.52	
F - 0% Ash and C.R.	1/06/2023	15/06/2023	14	39.50	12.00	19.80	34.00	490	5.31	**4.52**
F - 0% Ash and C.R.	1/06/2023	29/06/2023	28	39.40	11.90	19.80	34.00	530	5.79	
F - 0% Ash and C.R.	1/06/2023	29/06/2023	28	39.50	11.90	19.90	34.00	510	5.52	
F - 0% Ash and C.R.	1/06/2023	29/06/2023	28	39.50	12.00	19.70	34.00	560	6.13	
F - 0% Ash and C.R.	1/06/2023	29/06/2023	28	39.40	12.00	19.80	34.00	580	6.29	
F - 0% Ash and C.R.	1/06/2023	29/06/2023	28	39.50	12.00	19.90	34.00	490	5.26	**5.80**

Note: This table shows the addition of 0% RCBCACR, having a flexural strength of 5.80 kg/cm^2 at 28 days, being our standard sample.

Table 14

Flexural strength of each specimen with 5% addition of RCBCARC

№ of Witnesses	Description	Date Moulding	Date Breakage	Age (days)	Length (cm)	Width (cm)	Height (cm)	Free span between supports (cm)	Charge Kg	Mr (Kg/cm)^2	Average Mr (Kg/cm)^2
1	F - 5% Ash and R.C.	1/06/2023	8/06/2023	7	39.30	12.00	19.90	34.00	200	2.15	

2	F - 5% Ash and R.C.	1/06/2023	8/06/2023	7	39.50	11.90	19.80	34.00	180	1.97	
3	F - 5% Ash and R.C.	1/06/2023	8/06/2023	7	39.50	12.00	19.80	34.00	140	1.52	
4	F - 5% Ash and R.C.	1/06/2023	8/06/2023	7	39.40	12.00	19.70	34.00	210	2.30	
5	F - 5% Ash and R.C.	1/06/2023	8/06/2023	7	39.50	12.00	19.80	34.00	190	2.06	**2.00**
6	F - 5% Ash and R.C.	1/06/2023	15/06/2023	14	39.35	11.90	19.90	34.00	320	3.46	
7	F - 5% Ash and R.C.	1/06/2023	15/06/2023	14	39.50	12.00	19.90	34.00	270	2.90	
8	F - 5% Ash and R.C.	1/06/2023	15/06/2023	14	39.40	12.00	19.80	34.00	380	4.12	
9	F - 5% Ash and R.C.	1/06/2023	15/06/2023	14	39.50	11.90	19.80	34.00	300	3.28	
10	F - 5% Ash and R.C.	1/06/2023	15/06/2023	14	39.40	11.90	19.80	34.00	390	4.26	**3.60**
11	F - 5% Ash and R.C.	1/06/2023	29/06/2023	28	39.50	12.00	19.90	34.00	350	3.76	
12	F - 5% Ash and R.C.	1/06/2023	29/06/2023	28	39.40	12.00	19.80	34.00	480	5.20	
13	F - 5% Ash and R.C.	1/06/2023	29/06/2023	28	39.50	11.90	19.80	34.00	400	4.37	
14	F - 5% Ash and R.C.	1/06/2023	29/06/2023	28	39.50	11.90	19.70	34.00	420	4.64	
15	F - 5% Ash and R.C.	1/06/2023	29/06/2023	28	39.40	12.00	19.80	34.00	390	4.23	**4.44**

Note: This table shows the addition of 5% RCBCACR, having a flexural strength of 4.44 kg/cm^2 at 28 days, below the standard sample.

Flexural strength of concrete blocks with 0% addition of RCBCACR.

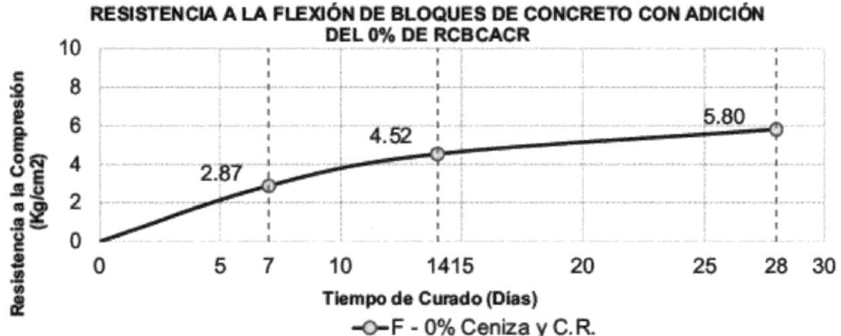

Note: This figure shows the addition of 0% RCBCACR, having a flexural strength of 5.80 kg/cm^2 at 28 days, used as the standard sample data.

Figure 17

Flexural strength of concrete blocks with 5% RCBCACR addition

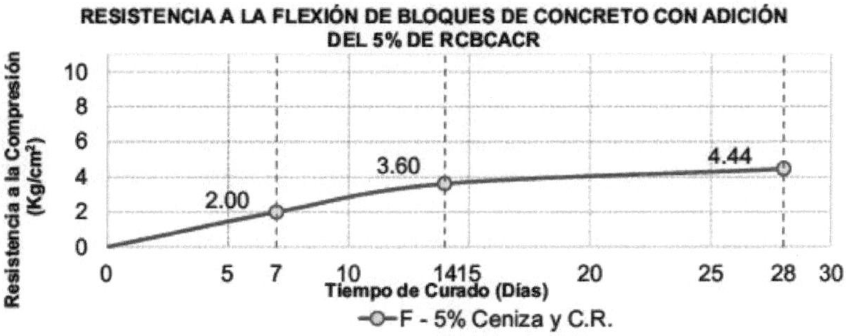

RESISTENCIA A LA FLEXIÓN DE BLOQUES DE CONCRETO CON ADICIÓN DEL 5% DE RCBCACR

Resistencia a la Compresión (Kg/cm²)

10 8 6 4 2 0

2.00 3.60 4.44

0 5 7 10 14 15 20 25 28 30

Tiempo de Curado (Días)

F - 5% Ceniza y C.R.

Note: This figure shows the addition of 5% RCBCACR, having a flexural strength of 4.44 kg/cm^2 at 28 days, below the standard sample.

Table 15

Flexural strength of each specimen with 10% addition of RCBCARC

№ of Witnesses	Description	Date Moulding	Date Breakage	Age (days)	Length (cm)	Width (cm)	Height (cm)	Free span between supports (cm)	Charge Kg	Mr. (Kg/cm)^2	Average Mr. (Kg/cm)^2
1	F -10% Ash and R.C.	1/06/2023	8/06/2023	7	39.60	11.90	19.90	34.00	280	3.03	
2	F -10% Ash and R.C.	1/06/2023	8/06/2023	7	39.40	11.90	19.80	34.00	230	2.51	
3	F -10% Ash and R.C.	1/06/2023	8/06/2023	7	39.40	11.80	19.80	34.00	290	3.20	
4	F -10% Ash and R.C.	1/06/2023	8/06/2023	7	39.60	11.90	19.90	34.00	300	3.25	
5	F -10% Ash and R.C.	1/06/2023	8/06/2023	7	39.50	12.00	19.90	34.00	360	3.86	**3.17**
6	F -10% Ash and R.C.	1/06/2023	15/06/2023	14	39.50	11.90	19.70	34.00	390	4.31	
7	F -10% Ash and R.C.	1/06/2023	15/06/2023	14	39.40	11.80	19.80	34.00	490	5.40	
8	F -10% Ash and R.C.	1/06/2023	15/06/2023	14	39.50	11.90	19.80	34.00	380	4.15	
9	F -10% Ash and R.C.	1/06/2023	15/06/2023	14	39.40	12.00	19.70	34.00	410	4.49	
10	F -10% Ash and C.R.	1/06/2023	15/06/2023	14	39.40	11.80	19.90	34.00	430	4.69	**4.61**
11	F -10% Ash and C.R.	1/06/2023	29/06/2023	28	39.60	11.80	19.90	34.00	550	6.00	
12	F -10% Ash and C.R.	1/06/2023	29/06/2023	28	39.40	12.00	19.80	34.00	580	6.29	
13	F -10% Ash and C.R.	1/06/2023	29/06/2023	28	39.40	11.90	19.80	34.00	490	5.36	
14	F -10% Ash and C.R.	1/06/2023	29/06/2023	28	39.50	12.00	19.90	34.00	650	6.98	
15	F -10% Ash and C.R.	1/06/2023	29/06/2023	28	39.50	12.00	19.90	34.00	590	6.33	**6.19**

Note: This table shows the addition of 10% RCBCACR, having a flexural strength of 6.19 kg/cm^2 at 28 days, being above the standard sample.

Table 16

Flexural strength of each specimen with 15% addition of RCBCARC

№ of Witnesses	Description	Date		Age (days)	Length (cm)	Width (cm)	Height (cm)	Free span between supports (cm)	Charge Kg	Mr. (Kg/cm)²	Average Mr. (Kg/cm)²
		Moulding	Breakage								
1	F -15% Ash and R.C.	1/06/2023	8/06/2023	7	39.50	11.90	19.90	34.00	190	2.06	
2	F -15% Ash and R.C.	1/06/2023	8/06/2023	7	39.60	11.90	19.80	34.00	370	4.04	
3	F -15% Ash and R.C.	1/06/2023	8/06/2023	7	39.50	12.00	19.80	34.00	130	1.41	
4	F -15% Ash and R.C.	1/06/2023	8/06/2023	7	39.50	11.90	19.80	34.00	210	2.30	
5	F -15% Ash and R.C.	1/06/2023	8/06/2023	7	39.50	12.00	19.90	34.00	140	1.50	**2.26**
6	F -15% Ash and R.C.	1/06/2023	15/06/2023	14	39.40	11.80	19.70	34.00	480	5.35	
7	F -15% Ash and R.C.	1/06/2023	15/06/2023	14	39.40	12.00	19.80	34.00	510	5.53	
8	F -15% Ash and R.C.	1/06/2023	15/06/2023	14	39.50	12.00	19.80	34.00	300	3.25	
9	F -15% Ash and R.C.	1/06/2023	15/06/2023	14	39.40	11.90	19.80	34.00	140	1.53	
10	F -15% Ash and R.C.	1/06/2023	15/06/2023	14	39.40	12.00	19.90	34.00	310	3.33	**3.80**
11	F -15% Ash and R.C.	1/06/2023	29/06/2023	28	39.50	12.00	19.70	34.00	450	4.93	
12	F -15% Ash and R.C.	1/06/2023	29/06/2023	28	39.40	11.90	19.70	34.00	530	5.85	
13	F -15% Ash and R.C.	1/06/2023	29/06/2023	28	39.60	12.00	19.90	34.00	410	4.40	
14	F -15% Ash and R.C.	1/06/2023	29/06/2023	28	39.50	11.90	19.80	34.00	480	5.25	
15	F -15% Ash and R.C.	1/06/2023	29/06/2023	28	39.40	11.90	19.80	34.00	520	5.68	**5.22**

Note: This table shows the addition of 15% RCBCACR, having a flexural strength of 5.22 kg/cm^2 at 28 days, being below the standard sample.

Flexural strength of concrete blocks with 10% addition of RCBCACR.

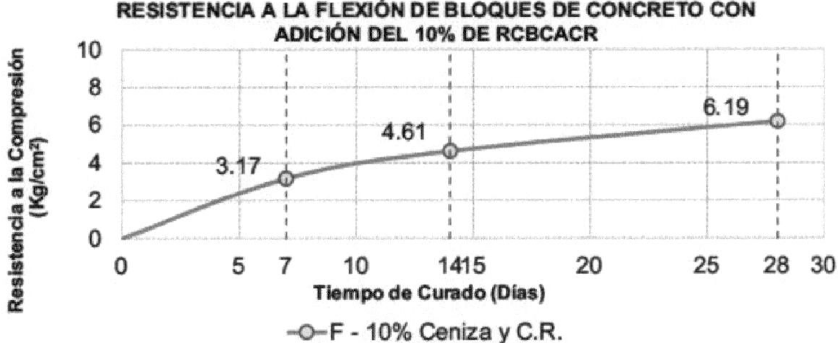

Note: This figure shows the addition of 10% RCBCACR, outperforms the other samples in strength, having a flexural strength of 6.19 kg/cm^2 at 28 days, above the standard sample.

Figure 19

Flexural strength of concrete blocks with 15% addition of RCBCACR.

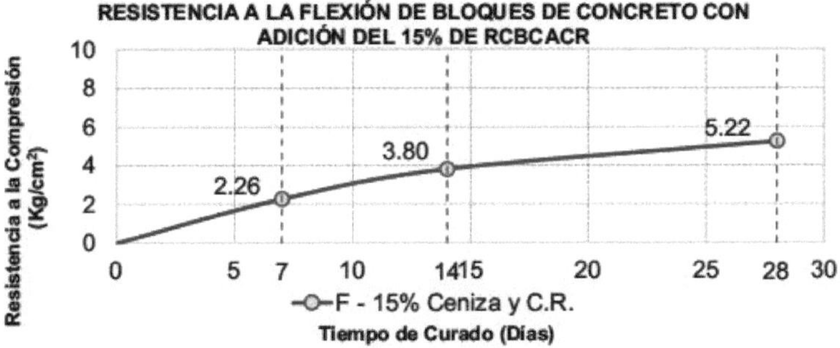

Note: This figure shows the addition of 15% RCBCACR, does not exceed the strength of the standard sample, having a flexural strength of 5.22 kg/cm^2 at 28 days.

Consolidated Flexural Strength of samples with 0%, 5%, 10% 15°/o addition of RCBCARC

Description	Date of breakage		Age (days)	Average Mr	Mr Design	%
	Moulding	Breakage		(Kg/cm)^2	(Kg/cm)^2	
F - 0% Ash and C.R.	1/06/2023	8/06/2023	7	2.9	-	-
F - 0% Ash and C.R.	1/06/2023	15/06/2023	14	4.5	-	-
F - 0% Ash and C.R.	1/06/2023	29/06/2023	28	5.8		
F - 5% Ash and R.C.	1/06/2023	8/06/2023	7	2.0	5.8	34%
F - 5% Ash and R.C.	1/06/2023	15/06/2023	14	3.6	5.8	62%
F - 5% Ash and R.C.	1/06/2023	29/06/2023	28	4.4	5.8	77%
F -10% Ash and C.R.	1/06/2023	8/06/2023	7	3.2	5.8	55%
F -10% Ash and R.C.	1/06/2023	15/06/2023	14	4.6	5.8	79%
F ▪ 10% Ash and R.C.	**1/06/2023**	**29/06/2023**	**28**	**6.19**	**5.8**	**107%**
F -15% Ash and R.C.	1/06/2023	8/06/2023	7	2.3	5.8	39%
F -15% Ash and R.C.	1/06/2023	15/06/2023	14	3.8	5.8	65%
F -15% Ash and R.C.	1/06/2023	29/06/2023	28	5.2	5.8	90%

Note: This table shows the final results of flexural strength with RCBCACR addition at 0%, 5%, 10% and 15%, for a batch of 5 samples, with the 10% dose being the best strength at 6.19 kg/cm .^{2}

Flexural strength of concrete blocks with addition of RCBCACR.

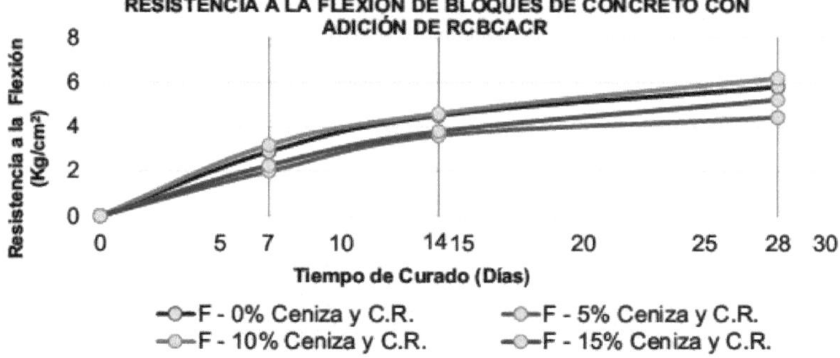

Note: This figure shows the final results of the flexural strength with RCBCACR addition at 0%, 5%, 10% and 15%, for a batch of 5 samples, with the 10% dose being the best strength with 6.19 kg/cm^2 , above the standard sample.

Table 18

Summary of flexural strength of concrete blocks with addition of RCBCACR.

BENDING STRENGTH (Kg/cm)^2			
Sample	**7 days**	**14 days**	**28 days**
0%	2.87	4.52	5.80
5%	2.00	3.60	4.44
10%	**3.17**	**4.61**	**6.19**
15%	2.26	3.80	5.22

Note: This table shows a summary of the flexural strength of the samples where the addition of 10% RCBCACR exceeds the strength with 6.19 kg/cm^2 at 28 days.

Flexural strength of concrete blocks with addition of RCBCACR.

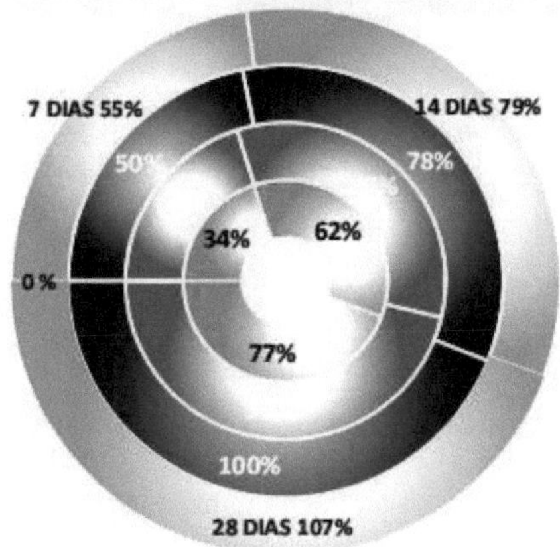

Note. This figure reflects the average results of the concrete blocks with addition of RCBCACR in 5. %, 10% and 15%, of the 3 groups, where the addition of 10% of RCBCACR at 28 days of age, is the dose that exceeds the standard in flexural strength, concluding with the general and specific hypothesis if they meet since the flexural strength improves significantly with the addition of RCBCACR, passing to 15% begins to decrease the resistance, similarly if it lowers to 5%, begins to lower the resistance.

Tensile strength of each sample with 0% addition of RCBCARC

Nº of Witnesses	Description	Date of breakage Moulding	Breakage	Age (days)	Length (cm)	Width (cm)	High (cm)	Perimeter Cut (cm)	Load (Kg)	Indirect Tensile Strength (kg/cm)^2	Average indirect tensile strength (kg/cm)^2	Design Strength % (kg/cm)^2
01	T -0% Ash and R.C.	2/06/2023	9/06/2023	7	39.50	12.00	19.80	63.60	450.00	1.18		
02	T -0% Ash and R.C.	2/06/2023	9/06/2023	7	39.40	11.90	19.90	63.60	470.00	1.24		
03	T -0% Ash and R.C.	2/06/2023	9/06/2023	7	39.40	12.00	19.70	63.40	520.00	1.37		
04	T -0% Ash and R.C.	2/06/2023	9/06/2023	7	39.50	11.90	19.80	63.40	490.00	1.30		
05	T -0% Ash and R.C.	2/06/2023	9/06/2023	7	39.40	11.90	19.90	63.60	460.00	1.22	1.19	
06	T -0% Ash and R.C.	2/06/2023	16/06/2023	14	39.30	12.00	19.80	63.60	450.00	1.18		
07	T -0% Ash and R.C.	2/06/2023	16/06/2023	14	39.50	11.00	19.90	61.80	450.00	1.32		
08	T -0% Ash and R.C.	2/06/2023	16/06/2023	14	39.50	9.00	19.90	57.80	420.00	1.61		
09	T -0% Ash and R.C.	2/06/2023	16/06/2023	14	39.40	12.00	19.80	63.60	430.00	1.13		
10	T -0% Ash	2/06/202	16/06/202	14	39.30	12.00	19.7	63.40	450.0	1.18	1.09	

№	Description	Moulding	Breakage	Age (days)	Length (cm)	Width (cm)	Height (cm)	Cut (cm)	Load (Kg)	Resistance to Indirect Traction (kg/cm)²	Average	Design
	and R.C.	3	3				0		0			
11	T -0% Ash and R.C.	2/06/2023	30/06/2023	28	39.50	11.90	19.90	63.60	400.00	1.06		
12	T -0% Ash and R.C.	2/06/2023	30/06/2023	28	39.40	11.80	19.80	63.20	430.00	1.15		
13	T -0% Ash and R.C.	2/06/2023	30/06/2023	28	39.40	12.00	19.80	63.60	450.00	1.18		
14	T -0% Ash and R.C.	2/06/2023	30/06/2023	28	39.50	12.00	19.70	63.40	380.00	1.00		
15	T -0% Ash and R.C.	2/06/2023	30/06/2023	28	39.30	11.90	19.80	63.40	450.00	1.19	**1.03**	

Note: This table shows the tensile strength with addition of RCBCACR at 0%, for a batch of 5 samples, being
our standard sample with a resistance of 1.03 kg/cm^2 in 28 days, in addition, it can be seen that the resistance has a
moderate decrease, this is due to the fact that the block does not work under traction.

Table 20

Tensile strength of each sample with 5% addition of RCBCARC

№ of Witnesses	Description	Date of breakage Moulding	Breakage	Age (days)	....Perimeter Length (cm)	Width (cm)	Height (cm)	Cut (cm)	Load (Kg)	Resistance to Indirect Traction (kg/cm)²	Resistance to Average Indirect Tensile Strength (kg/cm)²	Resistance Design (kg/cm)²	%
01	T - 5% Ash and C.R.	2/06/2023	9/06/2023	7	39.40	11.90	19.90	63.60	420.00	1.11			
02	T - 5% Ash and C.R.	2/06/2023	9/06/2023	7	39.50	12.00	19.90	63.80	410.00	1.07			
03	T - 5% Ash and C.R.	2/06/2023	9/06/2023	7	39.50	12.00	19.80	63.60	410.00	1.07			
04	T - 5% Ash and C.R.	2/06/2023	9/06/2023	7	39.50	12.00	19.80	63.60	380.00	1.00			
05	T - 5% Ash and C.R.	2/06/2023	9/06/2023	7	39.40	12.00	19.90	63.80	400.00	1.04	1.02	1.03	98.55 %
06	T - 5% Ash and C.R.	2/06/2023	16/06/2023	14	39.40	11.90	19.90	63.60	380.00	1.00			
07	T - 5% Ash and C.R.	2/06/2023	16/06/2023	14	39.50	11.90	19.80	63.40	410.00	1.09			
08	T - 5% Ash and C.R.	2/06/2023	16/06/2023	14	39.40	12.00	19.70	63.40	420.00	1.10			
09	T - 5% Ash and C.R.	2/06/2023	16/06/2023	14	39.50	12.00	19.90	63.80	400.00	1.04			
10	T - 5% Ash and	2/06/2023	16/06/2023	14	39.50	11.90	19.80	63.40	390.00	1.03	1.01	1.03	98.30 %

	C.R.												
11	T - 5% Ash and C.R.	2/06/2023	30/06/2023	28	39.40	12.00	19.80	63.60	410.00	1.07			
12	T - 5% Ash and C.R.	2/06/2023	30/06/2023	28	39.50	12.00	19.90	63.80	390.00	1.02			
13	T - 5% Ash and C.R.	2/06/2023	30/06/2023	28	39.50	11.90	19.90	63.60	320.00	0.85			
14	T - 5% Ash and C.R.	2/06/2023	30/06/2023	28	39.50	11.90	19.80	63.40	380.00	1.01			
15	T - 5% Ash and C.R.	2/06/2023	30/06/2023	28	39.60	12.00	19.90	63.80	400.00	1.04	**0.91**	1.03	88.10 %

Note: This table shows the tensile strength with the addition of RCBCACR at 5%, for a batch of 5 samples, this dosage decreases the strength to 0.91 kg/cm^2 in 28 days, not being appropriate.

Tensile strength of concrete blocks with 0% addition of RCBCACR.

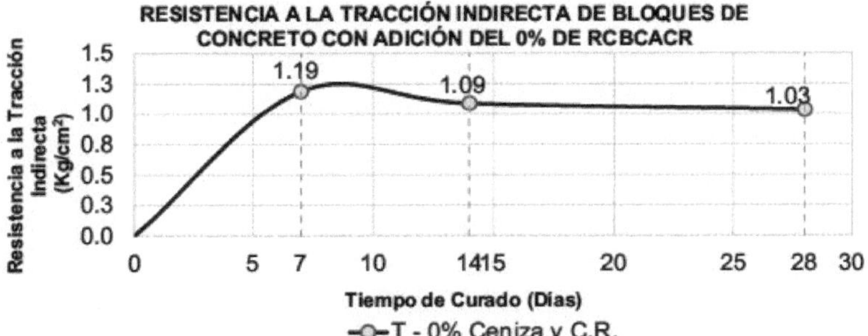

Note: This figure shows the addition of 0% RCBCACR, being our standard sample with a tensile strength of 1.03 kg/cm^2 at 28 days.

Figure 23

Tensile strength of concrete blocks with 5% addition of RCBCACR.

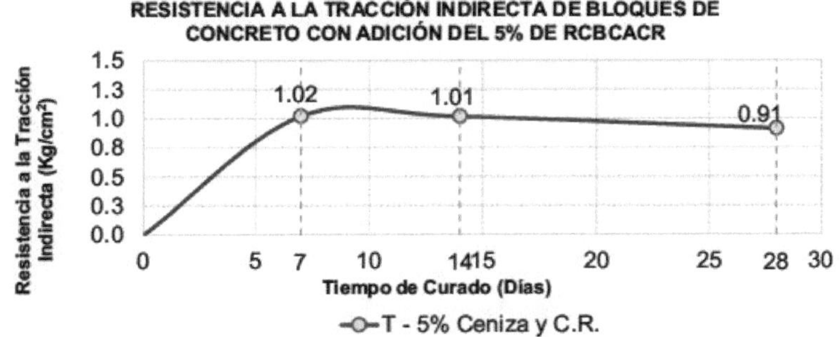

Note: This figure shows the addition of 5% RCBCACR, tending to lower its tensile strength by 0.91 kg/cm^2 at 28 days.

Tensile strength of each sample with 10% addition of RCBCARC

№ of Witnesses	Description	Date de Moulding	Date de Breakage	Age (days)	Length (cm)	Width (cm)	High (cm)	Perimeter Cut (cm)	Load (Kg)	Indirect Tensile Strength $(kg/cm)^2$	Average indirect tensile strength $(kg/cm)^2$	Design Strength $(kg/cm)^2$	%
01	T -10% Ash and R.C.	2/06/2023	9/06/2023	7	39.40	11.90	19.90	63.60	490.00	1.29			
02	T -10% Ash and R.C.	2/06/2023	9/06/2023	7	39.50	12.00	19.90	63.80	510.00	1.33			
03	T -10% Ash and R.C.	2/06/2023	9/06/2023	7	39.50	11.90	19.80	63.40	450.00	1.19			
04	T -10% Ash and R.C.	2/06/2023	9/06/2023	7	39.50	12.00	19.80	63.60	520.00	1.36			
05	T -10% Ash and R.C.	2/06/2023	9/06/2023	7	39.40	12.00	19.90	63.80	500.00	1.31	1.23	1.03	119.55 %
06	T -10% Ash and R.C.	2/06/2023	16/06/2023	14	39.40	11.90	19.90	63.60	420.00	1.11			
07	T -10% Ash and R.C.	2/06/2023	16/06/2023	14	39.50	11.90	19.80	63.40	460.00	1.22			
08	T -10% Ash and R.C.	2/06/2023	16/06/2023	14	39.40	12.00	19.70	63.40	470.00	1.24			
09	T -10% Ash and R.C.	2/06/2023	16/06/2023	14	39.50	12.00	19.90	63.80	440.00	1.15			
10	T -10% Ash and R.C.	2/06/2023	16/06/2023	14	39.50	11.90	19.80	63.40	490.00	1.30	1.13	1.03	109.35 %
11	T -10% Ash and R.C.	2/06/2023	30/06/2023	28	39.40	12.00	19.80	63.60	450.00	1.18			
12	T -10% Ash and R.C.	2/06/2023	30/06/2023	28	39.50	12.00	19.90	63.80	420.00	1.10			
13	T -10% Ash and R.C.	2/06/2023	30/06/2023	28	39.50	11.90	19.90	63.60	460.00	1.22			
14	T -10% Ash and R.C.	2/06/2023	30/06/2023	28	39.50	11.90	19.80	63.40	410.00	1.09			
15	T -10% Ash and R.C.	2/06/2023	30/06/2023	28	39.60	12.00	19.90	63.80	430.00	1.12	**1.09**	1.03	105.17 %

Note: This table shows the tensile strength with addition of RCBCACR at 10%, for a batch of 5 samples, with a strength of 1.09 kg/cm^2 in 28 days, it can be seen that the strength has a moderate decrease, but with a higher strength than the standard.

Tensile strength of each sample with 15% addition of RCBCARC

№ of Witnesses	Description	Date of breakage Moulding	Breakage	Age (days)	Length (cm)	Width (cm)	High (cm)	Perimeter Cut (cm)	Load (Kg)	Indirect Tensile Strength (kg/cm)^2	Average indirect tensile strength (kg/cm)^2	Design Strength (kg/cm)^2	%
01	T -15% Ash and R.C.	2/06/2023	9/06/2023	7	39.40	11.90	19.90	63.60	480.00	1.27			
02	T -15% Ash and R.C.	2/06/2023	9/06/2023	7	39.50	12.00	19.90	63.80	430.00	1.12			
03	T -15% Ash and R.C.	2/06/2023	9/06/2023	7	39.50	12.00	19.80	63.60	500.00	1.31			
04	T -15% Ash and R.C.	2/06/2023	9/06/2023	7	39.50	12.00	19.80	63.60	410.00	1.07			
05	T -15% Ash and R.C.	2/06/2023	9/06/2023	7	39.40	12.00	19.90	63.80	510.00	1.33	1.11	1.03	107.18 %
06	T -15% Ash and R.C.	2/06/2023	16/06/2023	14	39.40	11.90	19.90	63.60	460.00	1.22			
07	T -15% Ash and R.C.	2/06/2023	16/06/2023	14	39.50	12.00	19.80	63.60	430.00	1.13			
08	T -15% Ash and R.C.	2/06/2023	16/06/2023	14	39.40	12.00	19.70	63.40	390.00	1.03			
09	T -15% Ash and R.C.	2/06/2023	16/06/2023	14	39.50	12.00	19.90	63.80	420.00	1.10			
10	T -15% Ash and R.C.	2/06/2023	16/06/2023	14	39.50	11.90	19.80	63.40	410.00	1.09	1.04	1.03	100.87 %
11	T -15% Ash and R.C.	2/06/2023	30/06/2023	28	39.40	11.90	19.80	63.40	400.00	1.06			
12	T -15% Ash and R.C.	2/06/2023	30/06/2023	28	39.50	12.00	19.90	63.80	390.00	1.02			
13	T -15% Ash and R.C.	2/06/2023	30/06/2023	28	39.50	11.90	19.90	63.60	390.00	1.03			
14	T -15% Ash and R.C.	2/06/2023	30/06/2023	28	39.50	11.90	19.80	63.40	350.00	0.93			
15	T -15% Ash and R.C.	2/06/2023	30/06/2023	28	39.60	12.00	19.90	63.80	380.00	0.99	**0.96**	1.03	92.65 %

Note: This table shows the tensile strength with the addition of RCBCACR at 15%, for a batch of 5 samples, this dosage decreases the strength to 0.96 kg/cm^2 in 28 days, not appropriate.

Tensile strength of concrete blocks with 10% addition of RCBCACR.

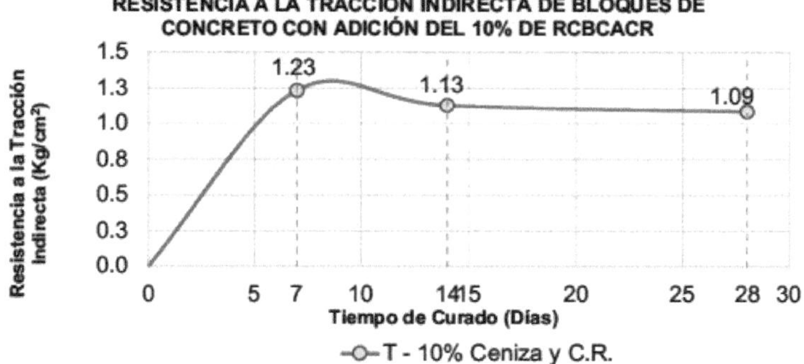

Note: This figure shows the addition of 10% RCBCACR, with a tensile strength of 1.09 kg/cm^2 at 28 days, with a strength higher than the standard.

Figure 25

Tensile strength of concrete blocks with 15% addition of RCBCACR.

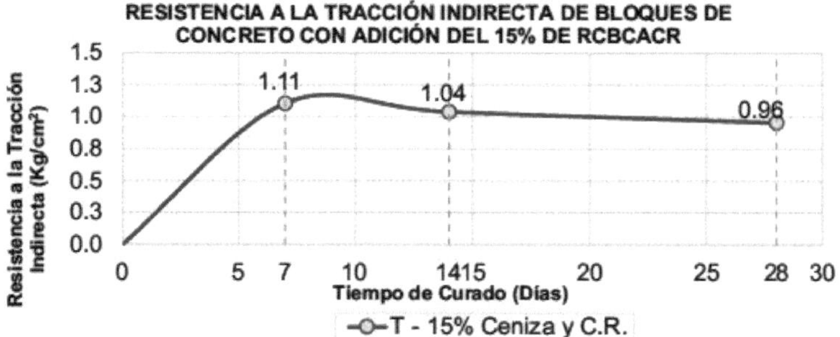

Note: This figure shows the addition of 15% RCBCACR, tending to lower its tensile strength by 0.96 kg/cm^2 at 28 days.

Consolidated Tensile Strength of samples with 0%, 5%, 10%, 15%> RCBCARC addition

Description	Date of breakage		Age (days)	Average indirect tensile strength (kg/cm)^2	Design Strength (kg/cm)^2	(%)
	Moulding	Breakage				
T - 0% Ash and C.R.	2/06/2023	9/06/2023	7.00	1.19	-	-
T - 0% Ash and C.R.	2/06/2023	16/06/2023	14.00	1.09	-	-
T - 0% Ash and C.R.	2/06/2023	30/06/2023	28.00	1.03	-	-
T - 5% Ash and C.R.	2/06/2023	9/06/2023	7	1.02	1.03	99%
T - 5% Ash and C.R.	2/06/2023	16/06/2023	14	1.01	1.03	98%
T - 5% Ash and C.R.	2/06/2023	30/06/2023	28	0.91	1.03	88%
T -10% Ash and R.C.	2/06/2023	9/06/2023	7	1.23	1.03	120%
T -10% Ash and R.C.	2/06/2023	16/06/2023	14	1.13	1.03	109%
T ■ 10% Ash and R.C.	**2/06/2023**	**30/06/2023**	**28**	**1.09**	**1.03**	**105%**
T -15% Ash and R.C.	2/06/2023	9/06/2023	7	1.11	1.03	107%

T -15% Ash and R.C.	2/06/2023	16/06/2023	14	1.04	1.03	101%
T -15% Ash and R.C.	2/06/2023	30/06/2023	28	0.96	1.03	93%

Note: This table shows the final results of the tensile strength with the addition of RCBCACR at 0%, 5%, 10% and 15%, for a batch of 5 samples, being the 10% dose the one with the best strength, although it is true that the strength is low because the blocks do not work in tensile, in relation to the other proportions its drop is lower with 1.09 kg/cm $.^2$

Tensile strength of concrete blocks with addition of RCBCACR.

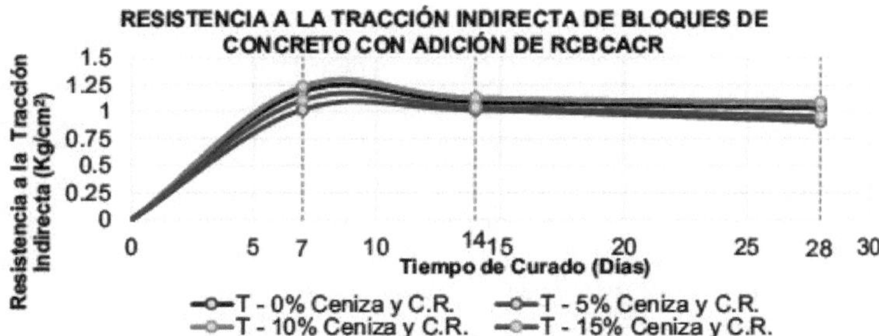

Note: This figure shows the final tensile strength results with the addition of RCBCACR at 0%, 5%, 10% and 15%, for a batch of 5 samples, with the 10% dose being the best strength at 1.09 kg/cm^2 , above the standard sample.

Table 24

Summary of tensile strength of concrete blocks with RCBCACR addition.

TENSILE STRENGTH (Kg/cm)^2			
Sample	**7 days**	**14 days**	**28 days**
0%	1.19	1.09	1.03
5%	1.02	1.01	0.91
10%	**1.23**	**1.13**	**1.09**
15%	1.11	1.04	0.96

Note: This table shows a summary of the tensile strength of the samples, where the addition of 10% RCBCACR, exceeds in strength with 1.03 kg/cm^2 at 28 days.

Tensile strength of concrete blocks with addition of RCBCACR.

% DE RESISTENCIA A LA TRACCION DE BLOQUES DE CONCRETO CON ADICIÓN DE RCBCACR

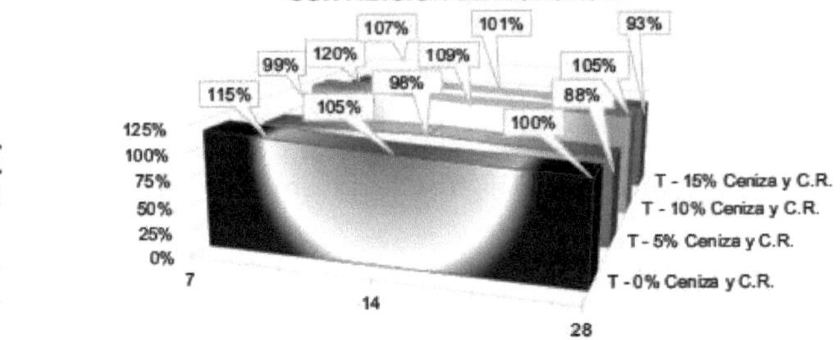

TIEMPO DE CURADO (DIAS)

T - 0% Ceniza y C.R. T - 5% Ceniza y C.R.
T - 10% Ceniza y C.R. T - 15% Ceniza y C.R.

Note: This figure reflects the average results of the concrete blocks with addition of RCBCACR in 5%, 10% and 15%, of the 3 groups, where the addition of 10% RCBCACR is the one that surpassed the standard in tensile strength at 28 days of age, concluding that the 10% addition of RCBCACR is the most optimal, if this percentage increases the strength decreases and if the percentage of addition decreases in the same way the tensile strength decreases.

Ratio of compressive strength Vs flexural strength of concrete blocks with RCBCACR addition.

Age Specimen	Dose	Average Compressive Strength (kg/cm2)	Average Flexural Strength (Kg/cm2) 10%F'c<=Mr<=20%F'c		
7		39.51	3.95	2.87	7.90
14		45.69	4.57	4.52	9.14
28	0%	**51.08**	**5.11**	**5.80**	**10.22**
7		38.46	3.85	2.00	7.69
14		38.91	3.89	3.60	7.78
28	5%	**42.58**	**4.26**	**4.44**	**8.52**
7		41.00	4.10	3.17	8.20
14		47.60	4.76	4.61	9.52
28	10%	**53.79**	**5.38**	**6.19**	**10.76**
7		38.76	3.88	2.26	7.75
14		40.80	4.08	3.80	8.16
28	15%	**46.23**	**4.62**	**5.22**	**9.25**

Note: This table shows a relationship that exists between compression and bending, Gerardo A. Rivera L. (2013); indicates that the relationship varies between 10% and 20% of the compressive strength, all three doses meet the condition, but the dose of 10% is higher with 6.19%.

Ratio of compressive strength Vs flexural strength of concrete blocks with RCBCACR addition.

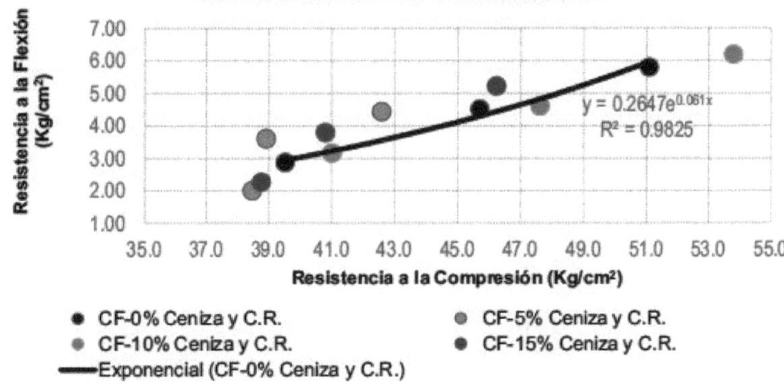

Note: This figure reflects the proximity of the 10% dosage to the standard sample, where the higher the compressive strength, the higher the flexural strength.

Ratio of compressive strength Vs flexural strength of RCBCACR added concrete blocks at 28 days.

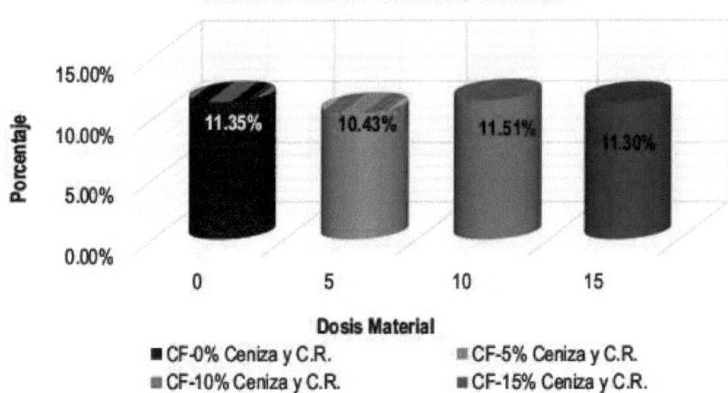

Note: This figure reflects the ratio of compressive strength vs. flexural strength of concrete blocks with addition of RCBCACR at 5%, 10% and 15%, its ratio of compressive strength vs. flexural strength, 10% addition is maintained as predominant, also if we take into account the average reference parameters for this study ranges between 10% - 20% (Gerardo A. Rivera L.), the ratio of compressive strength and flexural strength, which would indicate that the addition of RCBCACR at 10% would work with flexural stresses. Rivera L), the ratio of compressive strength to flexural strength, which would indicate that the addition of RCBCACR at 10% would work with flexural stresses.

Percentage ratio of compressive strength vs. tensile strength
Indirect concrete block with addition of RCBCACR at 28 days.

Age	Dose	Average	Average Indirect Tensile Strength

Specimen		Compressive Strength (kg/cm2)	(kg/cm2) 8%F'c<Mr<15%F'c		
7		39.51	3.16	1.19	5.93
14		45.69	3.66	1.09	6.85
28	0%	**51.08**	**4.09**	**1.03**	**7.66**
7		38.46	3.08	1.02	5.77
14		38.91	3.11	1.01	5.84
28	5%	**42.58**	**3.41**	**0.91**	**6.39**
7		41.00	3.28	1.23	6.15
14		47.60	3.81	1.13	7.14
28	10%	**53.79**	**4.30**	**1.09**	**8.07**
7		38.76	3.10	1.11	5.81
14		40.80	3.26	1.04	6.12
28	15%	**46.23**	**3.70**	**0.96**	**6.93**

Note: This table shows a relationship that exists between compression and traction, Otazzi (2004); indicates that the relationship varies between 8% and 15% of the compressive strength, which does not meet the three doses condition, but the dose of 10% is higher with 1.09%, this does not comply taking into account that the block does not work in traction.

Ratio of compressive strength Vs tensile strength of concrete blocks with addition of RCBCACR.

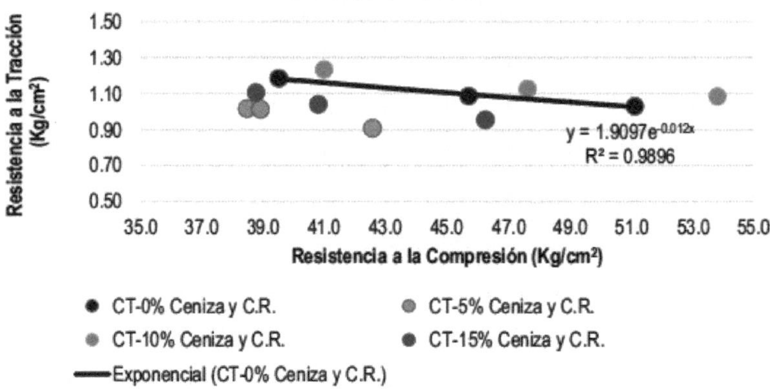

Note: This figure reflects the proximity of the 10% dosage to the standard sample, where the higher the compressive strength, the higher the indirect tensile strength.

Percentage Ratio of compressive strength Vs tensile strength of concrete blocks with addition of RCBCACR at 28 days.

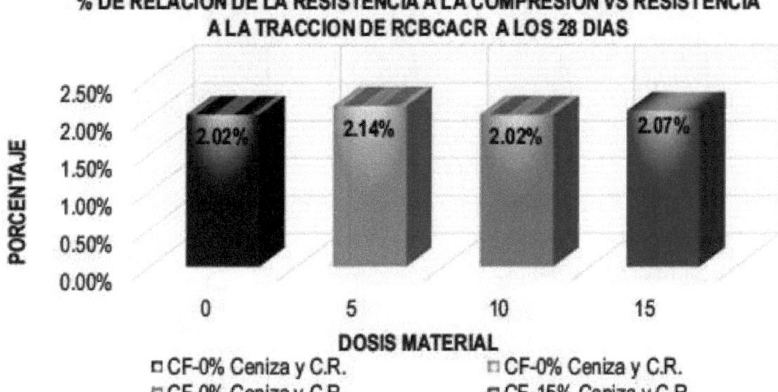

Note. The figure shows the ratio of compressive strength vs. indirect tensile strength of concrete blocks with addition of RCBCACR in 5%, 10% and 15%, shows a similar trend with respect to the standard sample, and if we take into account the average reference parameters for this study ranges between 8% - 15% (Farrel), the ratio of compressive strength vs. indirect tensile strength, these would stretch considerably below these parameters, indicating that the addition of RCBCACR to concrete blocks do not provide resistance to tensile stresses.

V.- DISCUSSION

V.1 In the present investigation it was obtained that the addition of RCBCACR to the concrete blocks in doses of 0%, 5%, 10% and 15% gives us as compressive strength of 51.08Kg/cm2, 42.58Kg/cm2, 53.79 Kg/cm^2 and 46.23 Kg/cm^2 , being the dose of 10% the most optimal in its compressive strength of 53.79 kg/cm^2 with respect to the other dosages, this is similar to the research conducted by Garrett et al. (2020). using 0%, 10%, 20% and 30% of sugar cane bagasse ash and rice husk, achieved a compressive strength of 18MPa, 25MPa, 17MPa and 18.5MPa for the first material and 20MPa, 25MPa and 24.5MPa for the second material. They concluded that 10% of sugar cane bagasse ash was the most suitable as it had significant results, for use in the production of concrete blocks.

V.2 Similarly, the present study has similar compressive strength results to the study carried out by Ardiles (2021) which presents results that by using 0, 5, 10 and 15 % of sugar cane bagasse ash, achieved compressive strengths of 28,180, 33,590, 37,810 and 35,450 kg/cm^2 , concluding that the most optimal dosage was 10% as it had a better value in terms of compressive strength.

V.3 In the research conducted by Prabhu et al. (2019) in which he made addition of bagasse ash in the manufacture of bricks. making use of 6, 8, 16 and 20 % achieving a compressive strength of 4.00N/mm^2 , 4.20N/mm^2 , 5.10N/mm^2 and 6.30N/mm^2 respectively, Concluding that by including 20% bagasse ash had significant results in its strength, so in the present investigation it is concluded the opposite by increasing the percentage of ash in concrete blocks from 10% upwards decreases its strength moderately.

V.4 According to the study carried out by Gerardo A. Rivera L. (2013), he concludes that the ratio of compressive strength vs. flexural strength ranges from 10% - 20%. In the present study, results were obtained for dosages of 0%, 5%, 10% and 15%, obtaining a ratio of compression vs. flexural strength of 11.35%, 10.43%, 11.51% and 11.30%, where the three dosages studied have results within the parameter, highlighting that with 10% of addition it has a higher strength compared to the other dosages.

It concludes that the ratio of compressive vs. flexural strength is closely related to the addition of RCBCACR.

V.5 According to a study by Otazzi (2004) the ratio of compressive strength vs. tensile strength is within the range of 8% - 15%, in the present study results of 2.02% - 2.14% were obtained, well below the reference range, to which we conclude that the addition of RCBCACR to concrete blocks does not provide tensile strength.

The methodology used in the design is adequate, because it allowed to determine the compressive, flexural and tensile strength, taking into account that the data analysis of the flexural and tensile strength has been an empirical comparison with the authors of the reference, since in our country there is still no standard that regulates flexural and tensile parameters for concrete block masonry units.

VI.-CONCLUSIONS

V.6 It was concluded that a nominal or standard design with a strength of f'_b = 50 kg/cm^2 at 28 days obtained an optimum design of f'_b = 51.08 kg/cm^2 at 28 days, an optimum design ratio for flexural strength of 11.35% at 28 days and an optimum design ratio for tensile strength of 2.02%, the standard concrete blocks were handmade which complies with the E.070-(RNE) standard.

V.7 The dosage of the addition of RCBCACR in percentages of 5%, 10% and 15%; allowing the manufacture of the sample units for the tests, obtaining: for the standard sample 0% addition - batch of 5 samples = cement 5.83 kg, ash 0 kg, water 5.01L, sand 40.29 kg, jam 4.43 kg, recycled concrete 0 kg; sample 5% addition - batch of 5 samples = cement 5.53 kg, ash 0.30 kg, water 5.01L, sand 40.29 kg, jam 4.20 kg, recycled concrete 0.23 kg; 10% addition sample - 5 sample batch = cement 5.23 kg, ash 0.60 kg, water 5.01 L., sand 40.29 kg, jam 4.20 kg, recycled concrete 0.23 kg; 10% addition sample - 5 sample batch = cement 5.23 kg, ash 0.60 kg, water 5.01L, sand 40.29 kg, whiting 3.97 kg, recycled concrete 0.46 kg; 15% addition sample - 5 sample batch = cement 4.93 kg, ash 0.90 kg, water 5.01L, sand 40.29 kg, whiting 3.74 kg, recycled concrete 0.69 kg; 15% addition sample - 5 sample batch = cement 4.93 kg, ash 0.90 kg, water 5.01L, sand 40.29 kg, whiting 3.74 kg, recycled concrete 0.69 kg.

V.8 The compressive strength of the concrete blocks with addition of RCBCACR in doses of 5%, 10% and 15% gives us 42.58Kg/cm^2 , 53.79 Kg/cm^2 and 46.23 Kg/cm^2 , having as a result that the doses of 5% and 15% give us lower strengths than the optimum design of the pattern (51.08 Kg/cm^2), so these units would only work as NP blocks, being the 10% dosage the most optimal in its compressive strength with 53.79 kg/cm^2 with respect to the other dosages; surpassing the standard sample at 28 days. Determining that with the 10% addition of RCBCACR better compressive strength is obtained, complying with Standard E.070-(RNE), being able to work as NP and P blocks.

V.9 The flexural strength of the concrete blocks with the addition of RCBCACR in dosages of 5%, 10% and 15% allowed us to conclude that the behaviour of RCBCACR acts in a favourable and proportional way in its three dosages of flexural strength (10.43%, 11.51% and 11.30%) at 28 days, highlighting that the 10% dosage of addition of RCBCACR has a higher ratio ($f'_{f/fb}$) with respect to its compressive strength. Determining that such concrete block units with 10% dosage resist better to flexural stresses, working as P and NP blocks.

V.10 The tensile strength of the concrete blocks with the addition of RCBCACR in dosages of 5%, 10% and 15%, concluding that the behaviour of RCBCACR acts unfavourably in its three dosages of tensile strength (2.14%, 2.02% and 2.07%) at 28 days, determining that these concrete block units do not resist tensile stresses.

VII.-RECOMMENDATIONS

According to the results obtained, the 10% addition of RCBCACR achieves a higher compressive strength (53.79 kg/cm^2), recommending the use of this dosage, since if the dosage is increased or decreased, its strength tends to decrease.

It is advisable to produce handmade concrete blocks, as long as the E.070-(RNE) standard is followed, and 28 days drying should be considered to reach an optimum strength.

According to the results obtained, as RCBCACRs are added or decreased, the compressive, flexural and tensile strength of the optimum dose obtained decreases towards both extremes, which is why the use of RCBCACRs at doses higher or lower than 10% is not recommended.

It is recommended not to perform the tensile strength test, as all tests performed act unfavourably.

In the present investigation, weights were taken into account, so it is recommended that future researchers conduct separate tests for RCBCACR in order to obtain more detailed information as to which of the two materials provides higher compressive, flexural and tensile strength.

REFERENCES

Abasi et al. (2023). EVALUATION of tensile properties of early-age concrete-block masonry assemblages. Retrievedfrom https://doi.org/10.1016/j.conbuildmat.2023.130542

Al-Aesh et al. (2021). Experimental assessment of the thermal and mechanical performance of insulated concrete blocks. Retrieved from https://doi.org/10.1016/j.jclepro.2020.124624

Almeida et al. (2019). USE of sugarcane bagasse ash sand (SBAS) as corrosion retardant for reinforced Portland slag cement concrete. Retrieved from https://doi.org/10.1016/j.conbuildmat.2019.07.217

Anjos et al. (2020). Properties of self-leveling mortars incorporating a high-volume of sugar cane bagasse ash as partial Portland cement replacement. Retrieved from https://doi.org/10.1016/j.jobe.2020.101694

ARDILES, R. (2021). *Influencia de la ceniza del bagazo de la caña de azúcar como sustituto parcial del cemento portland tipo I en la elaboración de unidades de albañilería Abancay, 2021.* UCV Institutional Digital Repository. Retrieved from https://hdl.handle.net/20.500.12692/75213

ARIAS, J. (2020). *Proyecto de tesis: Guía para la elaboración.* Retrieved from https://repositorio.concytec.gob.pe/bitstream/20.500.12390/2236/1/AriasGonzales_ProyectoDeTesis_libro.pdf

ARIAS, J., & COVINOS, M. (2021). *Research design and methodology.* Enfoques Consulting EIRL. Retrieved from https://repositorio.concytec.gob.pe/handle/20.500.12390/2260

Arivazhagan et al. (2020). Performance analysis of concrete block integrated with PCM for thermal management. Retrieved from https://doi.org/10.1016/j.matpr.2019.06.714

Awoyera et al. (2021). Water absorption, strength and microscale properties of interlocking concrete blocks made with plastic fibre and ceramic aggregates. Retrieved from https://doi.org/10.1016/j.cscm.2021.e00677

Cabané et al. (2022). Anisotropy and compressive strength evaluation of solid fired clay bricks by testing small specimens. Retrieved from https://doi.org/10.1016/j.conbuildmat.2022.128195

Camarena, A., & Díaz, D. (2022). *Comparative analysis of the compressive strength, flexural strength and workability of traditional concrete versus concrete using steel slag as fine aggregate.* Gaceta Técnica, 23 (1): 20-34. Retrieved from http://ve.scielo.org/scielo.php?script=sci_arttext&pid=S1856-95602022000100020

Carriço et al. (2021). Novel separation process for obtaining recycled cement and high- quality recycled sand from waste hardened concrete. Retrieved from https://doi.org/10.1016/j.jclepro.2021.127375

CAYOTOPA, K. (2019). *Compressive strength of concrete bricks f'c=210 kg/cm2, replacing the coarse aggregate by recycled brick and concrete, in different percentages.* UPN Institutional Repository. Retrieved from https://hdl.handle.net/11537/22301

CCAHUAYA, P., & ZEBALLOS, V. (2022). *Improvement of mechanical properties of the solid artisan wall by adding crushed recycled concrete to be used in construction, Ilo, Moquegua, 2021.* UCV Institutional Digital Repository. Retrieved from https://hdl.handle.net/20.500.12692/92011

CORREA, L., & POLO, H. (2019). *Influence of sugar cane ash replacement on the physical and mechanical properties of type II pavers for light traffic pavements, Trujillo 2019.* Trujillo: UPN Institutional Repository. Retrieved from https://hdl.handle.net/11537/23400

Dawoud et al. (2020). A Review on Investigating the experimental process for partial replacement of cement with sugarcane bagasse in the construction industry. *IOP Conference Series: Materials Science and Engineering*, 1-27. Retrieved from https://doi.org/10.1088/1757-899X/974/1/012036

Fakharian et al. (2023). Compressive strength prediction of hollow concrete masonry blocks using artificial intelligence algorithms. Retrieved from https://doi.org/10.1016/j.istruc.2022.12.007

Farrel et al. (1967). *Resistance Tests.* miscellaneous: miscellaneous.

Feng et al. (2023). Changing the soaking method of microbially induced calcium carbonate precipitation technology to improve the reinforcement effect of recycled concrete aggregates. Retrieved from https://doi.org/10.1016/j.jobe.2023.106128

Garrett et al. (2020). Sugarcane bagasse and rice husk ash pozzolans: Cement strength and corrosion effects when using saltwater. Retrieved from https://doi.org/10.1016/j.crgsc.2020.04.003

INEI. (2022). *National production.* Lima. Retrieved from https://www.inei.gob.pe/media/MenuRecursivo/boletines/03-informe-tecnico-produccion-nacional-ene-2022.pdf

Itamar et al. (2022). Systematic review of the literature on improving the mechanical properties of concrete with fibres of artificial-natural origin. Retrieved from https://doi.org/10.14483/23448393.18207

Khatab et al. (2021). The influence of waste of concrete masonry units as coarse aggregate on concrete properties. Retrieved from https://doi.org/10.1016/j.matpr.2020.12.186

Kolawole et al. (2021). State-of-the-art review on the use of sugarcane bagasse ash in cementitious materials. Retrieved from https://doi.org/10.1016/j.cemconcomp.2021.103975

L., G. A. (2013). *Simple concrete.* Cauca: E.

León, F., & Reátegui, S. (2020). *Design of concrete blocks incorporating sugar cane fibre for single-family houses in Moyobamba.* UCV Institutional Digital Repository. Retrieved from https://hdl.handle.net/20.500.12692/55360

Liu et al. (2020). Utilization potential of aerated concrete block powder and clay brick powder from C&D waste. Retrieved from https://doi.org/10.1016/j.conbuildmat.2019.117721

Liu et al. (2021). The potential use of drinking water sludge ash as supplementary cementitious material in the manufacture of concrete blocks. Retrieved from https://doi.org/10.1016/j.resconrec.2020.105291

Lyra et al. (2019). Reuse of sugarcane bagasse ash to produce a lightweight aggregate using microwave oven sintering. Retrieved from

https://doi.org/10.1016/j.conbuildmat.2019.06.150
MAKUL, N. (2020). *Cost-benefit analysis of the production of ready-mixed highperformance concrete made with recycled concrete aggregate: A case study in Thailand.* Heliyon, 6 (6): 1-13. Retrieved from https://doi.org/10.1016/j.heliyon.2020.e04135

Maldonado et al. (2019). Long -Term Corrosion Risk of Thin Cement Composites Containing Untreated Sugarcane Bagasse Ash. Retrieved from https://doi.org/10.1061/(ASCE)MT.1943-5533.0002647

MICHEAL, A., & MOUSSA, R. (2021). *Investigating the Economic and Environmental Effect of Integrating Sugarcane Bagasse (SCB) Fibers in Cement Bricks.* Ain Shams Engineering Journal, 12 (3): 3297-3303. Retrieved from https://doi.org/10.1016/j.asej.2020.12.012

Mucha et al. (2020). Evaluation of procedures used to determine the population and sample in postgraduate research. Retrieved from https://doi.org/10.37711/desafios.2021.12.1.253

NedeljkoviC et al. (2021). Use of fine recycled concrete aggregates in concrete: A critical review. Retrieved from https://doi.org/10.1016/j.jobe.2021.102196

Nikolenko et al. (2021). FLEXURAL strength of fibre reinforced concrete structures. Retrieved from https://doi.org/10.1088/1742-6596/1889/2/022075

NIÑO, V. (2021). *Research methodology.* Ediciones de la U. Retrieved from https://www.google.com.pe/books/edition/Metodolog%C3%ADa_de_la_investiga ci%C3%B3n/WCwaEAAAAQBAJ?hl=en-419&gbpv=0

Nunton et al. (2022). A review of the mechanical behaviour of concrete with the addition of steel fibres from recycled tyres. Retrieved from https://doi.org/10.25100/iyc.v24i2.11741

Ñaupas et al. (2019). Quantitative-qualitative research methodology and thesis writing. Retrieved from https://www.google.com.pe/books/edition/Metodolog%C3%ADa_de_la_Investiga ci%C3%B3n_cuanti/KzSjDwAAQBAJ?hl=en-419&gbpv=0

Pavlu et al. (2019). The utilization of recycled masonry aggregate and recycled EPS for concrete blocks for mortarless masonry. Retrieved from https://doi.org/10.3390%2Fma12121923

Peng et al. (2022). EXPERIMENTAL characterisation of masonry unit-mortar interface under uniaxial cyclic tension. Retrieved from https://doi.org/10.1016/j.engfracmech.2022.108790

Prabhu et al. (2019). An experimental study on bricks by partial replacement of bagasse ash. *International Research Journal of Multidisciplinary*, 258-265. Retrieved from https://doi.org/10.34256/irjmtcon35

RAMÍREZ, J., & CALLES, R. (2021). *Manual de metodología de la investigación en negocios internacionales.* Ecoe Ediciones. Retrieved from https://www.google.com.pe/books/edition/Manual_de_metodolog%C3%ADa_de _la_investigaci/GT4xEAAAAAQBAJ?hl=en-419&gbpv=0

REYES, E. (2022). *Methodology of Scientific Research.* Page Publishing, Incorporated. Retrieved from https://www.google.com.pe/books/edition/Metodologia_de_la_Investigacion_Cie ntifi/SmdxEAAAAAQBAJ?hl=en-419&gbpv=0

RNE (2019). *Standard E.070 Masonry.* Retrieved from https://www.cip.org.pe/publicaciones/2021/enero/portal/e.070-alba-ileria- sencico.pdf
Rodríguez et al. (2021). The variables in the methodology of scientific research. Retrieved from https://www.google.com.pe/books/edition/Las_variables_en_la_metodolog%C3%ADa_de_la_i/5jFJEAAAQBAJ?hl=en-419&gbpv=0
RODRÍGUEZ, Y. (2020). *Research methodology.* Klik Educational Solutions. Retrieved from https://www.google.com.pe/books/edition/Metodolog%C3%ADa_de_la_investigaci%C3%B3n/x9s6EAAAAAQBAJ?hl=en-419&gbpv=1
SÁNCHEZ, F. (2019). *Epistemic foundations of qualitative and quantitative research: Consensus and dissent.* Revista Digital de Investigación en Docencia Universitaria, 13(11): 102-122. Retrieved from http://dx.doi.org/10.19083/ridu.2019.644
SANCHEZ, N. (2011). *The management model and its impact on the provision of drinking water and sewerage services in the municipality of tena.* Ambato, Ecuador.
ßim§ek et al. (2022). Performance of fly ash-blended Portland cement concrete developed by using fine or coarse recycled concrete aggregate. Retrieved from https://doi.org/10.1016/j.conbuildmat.2022.129431
Song et al. (2019). Effect of carbon fiber on mechanical properties and dimensional stability of concrete incorporated with granulated-blast furnace slag. Retrieved from https://doi.org/10.1016/j.jclepro.2019.117819
Sun et al. (2020). Utilization of waste concrete recycling materials in self-compacting concrete. Retrieved from https://doi.org/10.1016/j.resconrec.2020.104930
Tayeh et al. (2021). Sugarcane pulp sand and paper grain sand as partial fine aggregate replacement in environment-friendly concrete bricks. *ELSEVIER.* Retrieved from https://doi.org/10.1016/j.cscm.2022.e01612
VILLANUEVA, F. (2022). *Research methodology.* Klil Soluciones Educativas. Retrieved from https://www.google.com.pe/books/edition/_/6e-KEAAAQBAJ?hl=es-419&gbpv=1
Wang et al. (2023). Study on the properties variation of recycled concrete paving block containing multiple waste materials. Retrieved from https://doi.org/10.1016/j.cscm.2022.e01803
Wu et al. (2022). Use of sugar cane bagasse ash in ultra-high performance concrete (UHPC) as cement replacement. Retrieved from https://doi.org/10.1016/j.conbuildmat.2021.125881
Yuan et al. (2023). MECHANICAL behavior and environmental assessment of steel-bars truss slab using steel fiber-reinforced recycled concrete. Retrieved from https://doi.org/10.1016/j.jobe.2023.106252
Zahra et al. (2021). Compressive strength and deformation characteristics of concrete block masonry made with different mortars, blocks and mortar beddings types. Retrieved from https://doi.org/10.1016/j.jobe.2021.102213
Zambrano et al. (2021). Application of curing methods and their influence on the compression strength of concrete. Retrieved from https://www.redalyc.org/journal/5703/570369777004/570369777004.pdf
https://miro.com/app/board/uXjVNU3_Wg4=/?share_link_id=521298544396

Printed by Books on Demand GmbH, Norderstedt / Germany